The Changing Role of the Embryo in Evolutionary Thought

In this book Ron Amundson examines 200 years of scientific views on the evolution–development relationship from the perspective of evolutionary developmental biology (evo–devo). This new perspective challenges several popular views about the history of evolutionary thought by claiming that many earlier authors made history come out right for the Evolutionary Synthesis.

The book starts with a revised history of nineteenth-century evolutionary thought. It then investigates how development became irrelevant to evolution with the Evolutionary Synthesis. It concludes with an examination of the contrasts that persist between mainstream evolutionary theory and evo–devo.

This book will appeal to students and professionals in the philosophy of science and the philosophy and history of biology.

Ron Amundson is Professor of Philosophy, University of Hawaii at Hilo.

CAMBRIDGE STUDIES IN PHILOSOPHY AND BIOLOGY

General Editor
Michael Ruse   *Florida State University*

Advisory Board
Michael Donoghue   *Yale University*
Jean Gayon   *University of Paris*
Jonathan Hodge   *University of Leeds*
Jane Maienschein   *Arizona State University*
Jesús Mosterín   *Instituto de Filosofía (Spanish Research Council)*
Elliott Sober   *University of Wisconsin*

Alfred I. Tauber   *The Immune Self: Theory or Metaphor?*
Elliott Sober   *From a Biological Point of View*
Robert Brandon   *Concepts and Methods in Evolutionary Biology*
Peter Godfrey-Smith   *Complexity and the Function of Mind in Nature*
William A. Rottschaefer   *The Biology and Psychology of Moral Agency*
Sahotra Sarkar   *Genetics and Reductionism*
Jean Gayon   *Darwinism's Struggle for Survival*
Jane Maienschein and Michael Ruse (eds.)   *Biology and the Foundation of Ethics*
Jack Wilson   *Biological Individuality*
Richard Creath and Jane Maienschein (eds.)   *Biology and Epistemology*
Alexander Rosenberg   *Darwinism in Philosophy, Social Science, and Policy*
Peter Beurton, Raphael Falk, and Hans-Jörg Rheinberger (eds.)   *The Concept of the Gene in Development and Evolution*
David Hull   *Science and Selection*
James G. Lennox   *Aristotle's Philosophy of Biology*
Marc Ereshefsky   *The Poverty of the Linnaean Hierarchy*
Kim Sterelny   *The Evolution of Agency and Other Essays*
William S. Cooper   *The Evolution of Reason*
Peter McLaughlin   *What Functions Explain*
Steven Hecht Orzack and Elliott Sober (eds.)   *Adaptationism and Optimality*
Bryan G. Norton   *Searching for Sustainability*
Sandra D. Mitchell   *Biological Complexity and Integrative Pluralism*
Joseph LaPorte   *Natural Kinds and Conceptual Change*
Greg Cooper   *The Science of the Struggle for Existence*
Jason Scott Robert   *Embryology, Epigenesis, and Evolution*
William F. Harms   *Information & Meaning in Evolutionary Processes*

The Changing Role
of the Embryo in
Evolutionary Thought

Roots of Evo–Devo

RON AMUNDSON

University of Hawaii at Hilo

CAMBRIDGE
UNIVERSITY PRESS

CAMBRIDGE UNIVERSITY PRESS
Cambridge, New York, Melbourne, Madrid, Cape Town, Singapore,
São Paulo, Delhi, Dubai, Tokyo, Mexico City

Cambridge University Press
The Edinburgh Building, Cambridge CB2 8RU, UK

Published in the United States of America by Cambridge University Press, New York

www.cambridge.org
Information on this title: www.cambridge.org/9780521703970

First published 2005
First paperback edition 2007

A catalogue record for this publication is available from the British Library

Library of Congress Cataloguing in Publication Data

Amundson, Ronald.
The changing role of the embryo in evolutionary thought : roots
of evo-devo / Ron Amundson.
    p. cm. – (Cambridge studies in philosophy and biology)
Includes bibliographical references and index.
ISBN 0-521-80699-2 (alk. paper)
1. Evolution (Biology) – Philosophy. 2. Developmental biology – Philosophy.
3. Embryology – Philosophy.   I. Title.   II. Series.
QH360.5.A48   2005
576.801 – dc22   2004054409

ISBN 978-0-521-80699-2 Hardback
ISBN 978-0-521-70397-0 Paperback

To my father, Art Amundson. He taught me curiosity, and the pleasure of taking things apart to see how they work.

Contents

ix

Acknowledgments

This book was written with the aid of a very large number of colleagues and friends. I owe special thanks to historian Polly Winsor, to biologists Scott Gilbert, Günter Wagner, and Brian Hall, and to philosophers Richard Burian and Rasmus Winther. Additional help and advice came from Gar Allen, Andrew Brook, Giovanni Camardi, Michael Dietrich, Raphael Falk, Frietson Galis, James Griesemer, Larry Heintz, Jane Maienschein, Staffan Müller-Wille, Lynn Nyhart, Rudy Raff, Marsha Richmond, Sara Scharf, David Wake, Sherry Amundson, the librarians of Mo'okini Library, and many others.

I live on the most remote island archipelago on earth. My physical impairments make overseas travel difficult, and everywhere is overseas from here. My research is successful only because of these colleagues' patience with my incessant emailed questions and my demands for obscure sources. (Don't get me wrong – I'm not complaining about life in Hawaii. I'm merely thanking the people who have helped to make it productive as well as pleasant.)

Since the mid-1980s I have also received generous support and advice from Ernst Mayr and David Hull, in full recognition of the fact that my conclusions were much at odds with theirs. I owe special thanks to my editor Michael Ruse, without whose blunt encouragement this book would never have been written. (Conversation in the beer garden at the 1995 Leuven meetings of ISHPSSB: "Do you mean to tell me you're about to turn fifty and you haven't written a book? You should be ashamed of yourself.")

Portions of Chapter 4 will appear as an introductory essay to a facsimile reprint of Richard Owen's *On the Nature of Limbs* (Owen 1849) forthcoming from the University of Chicago Press. Research was supported by the National Science Foundation under Grant SES-0135451.

1

Introduction

1.1 EVO–DEVO AS A NEW AND OLD SCIENCE

At the annual meeting of the Society for Integrative and Comparative Biology in January of the year 2000, a new Division was formed: the Division of Evolutionary Developmental Biology. This new organization would serve as a home for a lively field by the same name: evolutionary developmental biology, popularly known as evo–devo. In the minds of many of its practitioners (especially the more junior ones), evo–devo was new. It was a product of the explosive growth in knowledge about molecular developmental genetics during the 1990s. In a sense they were right; evo–devo really was new. Without the new molecular knowledge, evolutionary developmental biology would not have gathered the number of researchers or achieved the remarkable results that it could boast by the year 2000. Nevertheless, the subject is more than 150 years old. The conceptual connection between the development of an individual (ontogeny) and the evolution of a lineage (phylogeny) predates the 1859 publication of Darwin's *Origin of Species*. However, if evolutionary developmental biology is an old study, how could it be thought to be new in the year 2000?

The answer is that for most of the twentieth century only a minority of evolutionary biologists believed that ontogenetic development had *any relevance at all* to evolution. The Evolutionary Synthesis of the 1930s and 1940s established the mainstream of evolutionary biology (Mayr and Provine 1980). Population genetics was regarded as a causally adequate model of the evolutionary process. Natural selection was the sole direct-giving mechanism of evolutionary change, and the phenomena of macroevolution (patterns of evolution above the species level) were simply extrapolated from microevolution (natural selection within populations). The ontogenetic development of individual organisms had no place in this framework.

1

I am a philosopher and self-styled historian of biology. I am primarily interested in theoretical and methodological debates between scientific views, rather than in scientific theories themselves. I am interested in the recent revival of evolutionary developmental biology for a special reason: The evolutionary irrelevance of developmental biology was argued on the basis of philosophical, methodological, and sometimes even historical grounds during the mid-twentieth to late twentieth century. The basic concepts of evolutionary theory were said to preclude the relevance of development to evolution. These principles were described, examined, and (mostly) approved by philosophers as well as scientists. They were used by historians and scientists in reporting the history of evolutionary biology both before and after Darwin. Narratives of the history of biology depicted the predecessors of today's evo–devo practitioners as metaphysically confused and scientifically regressive. Pre-Darwinian biology was described in ways that detracted from the importance of developmental thinkers and that categorized them, along with almost all other opponents of Darwin, as religious reactionaries.

In other words, many philosophers and historians during the mid-twentieth to late twentieth century produced work that showed neo-Darwinism in a favorable light and developmental evolutionary theories in an unfavorable light. This is perfectly understandable, and I would have it no other way. Philosophers of science ought to take contemporary scientific knowledge as their starting point, and they ought not to feign wisdom that is superior to that of their scientific colleagues. In fact, I intend to do the very same thing in this book. I intend to look at the history and philosophy of biology *from the standpoint* of contemporary science. However, I will take a different standpoint from those who assumed the adequacy of the Evolutionary Synthesis.

Nothing succeeds like success. Evo–devo is a flourishing enterprise, notwithstanding the arguments and historical narratives of earlier days. I climb on this bandwagon here. I conduct my philosophical and historical examination from the standpoint of evo–devo rather than the Evolutionary Synthesis. Thus, the difference between this book and writings associated with the Evolutionary Synthesis is that this book has a *different* vantage point, a vantage point that has gained new legitimacy from recent science.

This book assumes the basic legitimacy of evo–devo. It examines certain traditional narratives of nineteenth-century biology with a view toward identifying and replacing the biases that made neo-Darwinian theory seem inevitable and alternative (especially developmental) theories seem regressive. It then examines the history of the twentieth-century interactions between evolutionary and developmental biology. Why was developmental biology absent from the early versions of neo-Darwinism? Why was it not later incorporated? Most importantly, what brought about the historical narratives and philosophical arguments that implied that development was *in principle irrelevant* to evolutionary biology?

It may seem that I am starting with a controversial assumption, that evo–devo and neo-Darwinism really are inconsistent. Surely they are not... well, probably they are not. Very few evo–devo practitioners doubt that natural selection within populations is responsible for the changes that occur within species. Evo–devo advocates merely believe that additional mechanisms, mechanisms involved with ontogeny rather than population genetics, must contribute to a full understanding of evolution. The problem is that the arguments constructed by neo-Darwinians that imply the irrelevance of development to (neo-Darwinian) evolution are very convincing! They entail that one can accept either evo–devo or neo-Darwinism, but not both; thus it is not my words, but the words of the neo-Darwinian commentators, that entail the inconsistency of evo–devo and neo-Darwinism. I hope, and most evo–devo practitioners believe, that a way can be found to accommodate both evo–devo and neo-Darwinism. There is a genuine tension between these viewpoints. I do not know how to refute the irrelevancy arguments of the neo-Darwinians. It is not yet clear how this dilemma will be resolved.[1]

Some readers will doubt that neo-Darwinians actually argued that development is irrelevant to the understanding of evolution, or that those arguments apply equally well to modern evo–devo. I document both assertions and do my best to explicate the tensions between the two views of evolution. I must leave it to others to resolve the tensions.

---

[1] Frankly, many evo–devo practitioners are not aware of these tensions. Most are aware of the practical barriers between the fields, such as the reliance of evo–devo on a relatively small number of model organisms and the lack of population-level studies. There are a range of opinions within the discipline regarding its relation to neo-Darwinian theory. Some practitioners, such as Brian Hall, consider evo–devo to be a new synthetic field of study that has no particular conflict with neo-Darwinism (Hall 2000). I discuss the contrast between Hall's own approach and that of neo-Darwinism in Chapter 11. Others recognize the conflicts but are optimistic about their resolution (Gilbert 2003b). One valuable approach to the history of evo–devo is to recognize its agenda, the contrast with the agenda of neo-Darwinism, and the various scientific disciplines that kept the evo–devo agenda alive during the twentieth century. These include comparative and experimental embryology, morphology, and paleontology (Love and Raff 2003). In this book I am primarily motivated by the specific methodological arguments that arose around 1980 concerning whether or not development was relevant to the understanding of evolution, and the philosophical and historical doctrines that gave rise to those arguments.

1.2 EVO–DEVO AND THE WINDFALL OF THE 1990s

Most nineteenth-century evolutionists and several twentieth-century evolutionists have argued for the importance of the processes of development in understanding evolution. These early views receive more attention in this book than the stunning molecular discoveries that stimulated the growth of evo–devo in the 1990s. I now briefly report on some of the discoveries of the 1990s to illustrate how new life was breathed into evolutionary developmental biology.

The 1970s and 1980s saw a number of iconoclastic challenges to the well-established Evolutionary Synthesis. Some of the criticisms have since been dropped (e.g., the alleged unfalsifiability of adaptationism), and some have become internal matters within mainstream evolutionary theory (e.g., the punctuation vs. gradualism issue in paleontology). The role of development in evolution is the single persistent dispute. It first took the form of an argument over "adaptation versus developmental constraints" (Maynard Smith et al. 1985; Amundson 1994; Schwenk 1995). That debate will be discussed later. For present purposes, the debate was important because it raised awareness of the significance (for the prodevelopment side) of the concept of homology. This new interest in homology coincided with the discovery by molecular biologists that protein molecules could be sequenced, and the similarity of sequences of different protein molecules could be measured. Like traditional anatomical homology, these molecular "homologies" could be compared in two ways: different forms of a certain category of protein within an individual (e.g., α – and β – globin molecules) is similar to anatomical "serial homology," and comparison between corresponding proteins in two species reveals "special homology." Like anatomical special homologies, closeness of match of molecular cross-species homologies was correlated with evolutionary relatedness. The serial homologies strongly suggest an evolutionary scenario in which the genetic basis of a single original protein had duplicated in some ancestor's genome, after which the duplicates independently diverged. Even these early molecular discoveries showed an intriguing similarity between nineteenth-century morphology and modern molecular biology (Gilbert 1980). Nothing radical is implied; both serially and specially homologous proteins merely exhibit evolutionary divergence.[2]

The molecular homologies among globin molecules were not at the time seen as developmental phenomena. The globin genes did not instruct development; rather their activation was seen as the consequence of the

---

[2] The brief narrative in this section follows Gilbert, Opitz, and Raff (1996).

4

interactions that caused certain cells to become red blood cells. They were the endpoint of differentiation, not its cause. Developmental implications began to take shape when the molecular techniques began to be applied to the genes that controlled the nature of specialized insect segments. The genetic experiments were inspired by homeotic mutations, a class of mutations discovered early in the twentieth century in which an insect segment, together with its ordinary appendages, was transformed into another type of segment; a *Drosophila* haltere could be transformed into a wing, or an antenna into a leg. These had been favorites of developmental evolutionary theorists such as William Bateson and Richard Goldschmidt. It was first discovered that the genes that produced the various homeotic mutations in *Drosophila* were themselves serially homologous. Moreover, they were located tandemly on a small region of a particular chromosome, and they were expressed on that chromosome in the same sequence as along the anterior–posterior axis in a fly's body. Each of these gene sequences contained a certain DNA sequence called the *homeobox*. These homeobox-containing genes came to be called *Hox* genes. So far so good. We were learning about the developmental genetics of *Drosophila* by identifying the genes that encode the proteins that determine segment identity.

The excitement really started when genes homologous to insect *Hox* genes were found in vertebrates. Insects and vertebrates are both segmented, but no one for the past century had seriously argued that segmentation was homologous between the two phyla. Then vertebrate genes similar in sequence to fly *Hox* genes were isolated. They proved to be arranged in the same order on the chromosome, and they were expressed in the same order in the body as the insect genes. "And last, it was shown that the enhancer region of a human homeotic gene, such as *deformed*, can function within *Drosophila* to activate gene expression in the same relative position as in the human embryo – in the head" (Gilbert et al. 1996: 364). Genes that act during development in a human's head can do their usual job in a developing fly's head. This was only the beginning of a sequence of shocking genetic homologies – homologies that firmly demonstrated phylogenetic relationships between groups whose anatomical characters almost no one had been so bold to identify as homologies. For example, the development of both the insect eye and the vertebrate eye is begun by the expression of homologous genes. The same is true with the hearts of insects and vertebrates, and with the limbs not only of insects and vertebrates but almost all other metazoan groups. More and more basic (and often analogous) body parts in diverse groups of organisms were found to be triggered by homologous genes. The implications are very hard to sort out, of course. Anatomical homologies have traditionally been identified either by their patterns of connections with other body parts or by their embryological origins. These initiating causes do not necessarily make the anatomical structures homologous (although they certainly challenge the traditional concepts of homology). Insect and vertebrate eyes are developed and structured in extremely different ways, even though they are the same with respect to the gene that begins their development. The difficult job for developmental genetics remains to show how the corresponding genes could serve as the original developmental triggers for such structurally distinct body parts. Tracing the genetic pathways and interactions "downstream" toward the eventual adult body part is an ongoing process; surprising new commonalities are revealed at every step.

These discoveries hearken to bygone days, and many developmental biologists knew it. One of the wildest homological speculations in history was put forth by Étienne Geoffroy St.-Hilaire in the 1820s. Geoffroy proposed that arthropods and vertebrates had identical body plans. The obvious problem (to knowledgeable anatomists) was that arthropods have their circulatory (haemal) system on their dorsal side and their neural system on their ventral side. Vertebrates are the reverse, with their neural spine along their back. This forced Geoffroy to suggest that the "identical" body plans were flipped upside -down with respect to the dorsal–ventral axis. Vertebrates travel with their neural spine toward the sun, whereas arthropods travel with their neural spine toward the earth. There was laughter all around. Toby Appel's 1987 book *The Cuvier–Geoffroy Debate* is quite sympathetic to Geoffroy. Still, she describes the arthropod–vertebrate body plan reversal as "preposterous," and she assures the reader that "such comparisons seemed no less fanciful to his contemporaries than they appear to us today" (Appel 1987: 111).

Geoffroy may have had the last laugh. Seven years after Appel's publication, it was discovered that the dorsal–ventral axes of vertebrates and arthropods are determined by homologous genes – but that their expression patterns were reversed in the two groups (De Robertis and Sasai 1996). Indeed, the expression patterns of an entire suite of genes used to specify the dorsal and ventral structures were inverted. These discoveries were not business as usual. Commonalities of animal structure that had previously been regarded as starry-eyed speculation were suddenly being traced to their molecular genetic roots.

The details of modern molecular developmental genetics are much more complex and fast changing than can be described here (Morange 1998). Genes are identified not in terms of the phenotypic effects that they produce in the adult. They are rather defined in terms of their roles in a "genetic toolkit" that is used, in different ways, in the embryological construction of the bodies of different kinds of organisms. The *Hox* system operates quite differently in insects and in vertebrates, but it operates in largely the same manner within the groups. The basic aspects of organic form are attributed to similar developmental processes, employing homologically similar developmental genes, or to similar "tools" (e.g., genetic processes). The repeated use of not simply genes but also genetic pathways has caused Scott Gilbert to speak of "homologies of process" rather than traditional anatomical homologies (Gilbert and Faber 1996; Gilbert and Bolker 2001). Attempts to understand how identical developmental genes can produce such diversity have led to an interpretation of developmental gene interactions as a kind of circuit, and major evolutionary changes as matters of the "rewiring" of genetic networks (Carroll, Grenier, and Weatherbee 2001; Wray 2001). Diversity is created by different applications of the same old tools. By *applications*, I mean the use of the same genetic systems in the actual building of the individual bodies of organisms of incredible diversity. This diversity is the product of the varying applications of shared developmental processes. Evo–devo itself goes well beyond the discoveries of deep homologies. It constructs evolutionary explanations; it doesn't just discover developmental–genetic causes. Most of the evo–devo explanations are consistent in spirit with developmental theories of past years. The dramatic new genetic homologies count as promises that there is much yet to be discovered. I discuss some of the evo–devo explanations, and their historical predecessors, later in the book. The dramatic new genetic homologies themselves will play no further role. I examine historical arguments, not modern discoveries.

Why are these new discoveries a problem for the neo-Darwinian critique of development? By announcing these dramatic discoveries at the beginning of this book, I may have made it difficult for the reader to imagine how anyone could doubt the importance of development to evolution. However, neo-Darwinism had its origins not in *developmental* genetics but in *transmission* genetics. Transmission genetics identifies individual genes not by their molecular sequence but by tracking phenotypic features through generations of organisms in breeding experiments. Genes are hypothesized on Mendelian principles in order to account for the patterns of the phenotypic features in offspring generations. The genes of transmission genetics are designed to explain the sorting of traits through generations; they expressly do not explain how traits are ontogenetically created within the individual organism. Population genetics, at the core of neo-Darwinian evolutionary theory, requires transmission genetics alone. It has absolutely no need for developmental genetics. This fact, together with the neo-Darwinian evolutionists' dislike for the developmental theorizing of the time, led to the antidevelopmental arguments.

And the arguments made sense. If populational processes are the only "mechanism" of evolutionary change, what difference does it make that human eyes and insect eyes originate from expression of the same gene?

The difference has to do with the significance of homology. As we will see, Darwin and his twentieth-century followers treat homology as a mere by-product of past evolutionary change, the leftover residue of ancestral characters that have not (yet) been selected out of the lineage. Homologies give evidence of past ancestry, but they are causally inert. Developmental evolutionists treat homology as an indicator of underlying causal processes of development that continue to exert their effects in contemporary species. These processes are the constraints in the "adaptation versus constraints" debates. The importance of the discovery of the deep genetic homologies is not just that one more homology has been detected. The discoveries were very special ones. The new deep homologies are causally active in the development of bodies, and that fact cannot be doubted. They are not mere residue. The very different bodies that are built by these genetic processes still show deep commonalities. Even the bilateral symmetry that characterizes such a wide variety of animal groups is no longer regarded as merely an efficient way to build bodies. It is a developmental heritage from an ancient common ancestor: *Urbilateria.*

The widespread sharing of developmentally important genes justifies a central assertion of evo–devo. It is that one must understand how bodies are built in order to understand how *the process of building bodies can be changed*, that is, how evolution can occur. The same arguments have been made since the early nineteenth century. The new genetic homologies offer new evidence that evolution cannot be understood without understanding development.

I examine the difference between Darwinian and developmental views of evolution during the course of this book. The book shows how an evo–devo sensibility produces a different narrative of the history of biology than a neo-Darwinian sensibility. I could not have written this book in 1990, prior to the discoveries of deep genetic homologies. The reason is not that my own arguments and historical narratives rely on the molecular discoveries themselves. They do not. The reason is that I intend to *assume* the legitimacy of evo–devo. I do not intend to argue for it. Such an assumption would have been controversial in 1990. The deep genetic discoveries allow me the same luxury that the neo-Darwinian commentators had between 1959 and the 1970s, when the philosophical and historical stage was being set. Like them, I can now reasonably assume that my favorite theory pretty much tells it like it is.

1.3 HOW I CAME TO THIS BOOK

I began studying these debates in the early 1980s, in the midst of the anti-Synthesis criticisms. I was just finishing an extended historical study of methodological conflicts in the history of experimental psychology between behaviorist and early cognitive psychologists (Amundson 1983, 1985, 1986). The two sides often seemed to argue past one another in these debates. However, I found that it was possible to discover hidden methodological conflicts by a close reading of the argumentation. Some features of the evolutionary debates of the 1970s and 1980s seemed very similar to me, especially those centering on development.[3] Adaptation versus developmental constraint was a function-versus-structure debate. The proadaptation side favored function over structure, and the prodevelopmental side favored structure (constraint) over function. I had just worked through a similar debate in psychology: the cognitivists were structuralists and the behaviorists were functionalists (Amundson 1989). I began reading in the history of evolutionary biology to see how deeply the structure–function contrast could be traced. It ran very deep indeed (Russell 1916). It seemed likely that the conflict between adaptation and developmental constraint was not only a phenomenon of the 1970s and 1980s.

My reading in the history of evolutionary biology has been guided by secondary historical sources. As I read through reports about pre-Darwinian British naturalists, I began to get the feeling that the deck had been stacked. Even in the secondary literature I could recognize structure–function debates between pre-Darwinian scientists. Their disagreements paralleled those of the 1980s. However, most historical commentaries failed to take that distinction seriously. They classified all pre-Darwinians into a single category of antievolutionists, and they glossed over the differences between functionalists and structuralists. This was my first hint that an examination of the methodological debates of the 1980s would extend into an examination of how the history of evolutionary biology had been written. Important pre-Darwinian conflicts had been historiographically minimized in a way that obscured the parallel between the pre-Darwinian structure–function debates and those of the 1980s.

---

[3] A sabbatical year in 1985–1986 spent in Stephen Jay Gould's lab, and regular discussions with Pere Alberch and Richard Lewontin, aided these thoughts.

To top it all off, the conceptual errors that were attributed to the (generic) pre-Darwinians were *exactly those conceptual errors* being attributed to the modern-day structuralist critics of the Synthesis!

"The game is afoot," thought I. "Someone is cooking the books!" (though perhaps not in those very words).

This was my first evidence that many histories of evolutionary biology had been written by people who considered the Evolutionary Synthesis to be essentially correct about evolutionary biology, including its opposition to modern alternative theories that involve development. The commitment to a particular modern theory had colored the reportage of historical science. Historical narratives could be read simultaneously as explanations of Darwin's 100-year-old success over his critics, and of the parallel success of the Evolutionary Synthesis over its modern critics. I realized that a historian who took the "constraints" side of the modern adaptation–constraints debate would write a very different history of evolutionary biology.

This is that history. I have cooked my own book.

I have since come to understand that writing the history of science is seldom an objective facts-only report of events. Scientists, especially when writing about the history of their own science, are simultaneously conducting contemporary research and argumentation. This is true of philosophers as well, who often have philosophical as well as scientific theories in the backs of their minds. Historians (especially recently) are somewhat less influenced by modern science, apparently because their discipline has provided them with other frameworks for their studies (e.g., the influence of social institutions or the self-interests of scientists on the practice of science). However, as we will see, historians too have a tendency to provide narratives that "come out right." A narrative *comes out right* when the predecessors of approved modern theories appear (in the narrative) to have made more sense than their contemporaries who turned out to be predecessors of theories that are now regarded as fallacious.

I will not attempt to avoid this problem of bias, but I will try to make it as transparent as possible. We (philosophers especially) do not do history from an abstract love of history. David Hull and I have come to quite different conclusions in our historical writings. In a discussion of our differences, David pointed out to me that my own writing was as biased as I claimed the traditional Synthesis histories to have been. He said that his work of the 1960s and 1970s was "history done in a good cause." He made me realize that mine is exactly the same. But it's now thirty years later. His good cause was won (with his able help), and my good cause is a different cause.

I had originally conceived this book as a chronological history of the relations between development and evolution. The chronological sequence has been roughly maintained, but the book is now separated into two parts that are distinct in methodology and in the centuries they cover. The two parts could very well be read separately. Although both parts are historical narratives, they are set against different backgrounds. In an odd way, the first part of the book is dependent on the second part.

The first part covers roughly the nineteenth century. The narrative is self-consciously revisionist; it is set against the background of traditional narratives that grew up with the Evolutionary Synthesis. I refer to the traditional narratives as Synthesis Historiography, or SH. I try to show that developmental approaches to evolution were scientifically progressive before Darwin, that they benefited Darwin's program, and that Darwin recognized that fact. In addition, the program of evolutionary morphology that immediately followed Darwin was itself well motivated and reasonable. It failed, but not because of the ideological flaws alleged by SH authors. Understanding the nineteenth century from this point of view, there is nothing that would lead one to believe that development is irrelevant to evolution except the practical difficulties of understanding how embryogenesis actually works. This is all contrary to the tradition of SH, which finds ideological flaws and metaphysical errors in every nineteenth-century advocate of developmental evolution.

The second part of the book covers the twentieth century. It is not particularly revisionist. Indeed, it supports many of the claims of SH authors (e.g., Ernst Mayr's claim that embryologists were not originally interested in participating in the Synthesis). The goal of the second part is to understand how the Evolutionary Synthesis *came to be opposed* to development in evolution – how things changed so very much from the nineteenth century. A part of this narrative is how SH (and its associated philosophical arguments) came into being. The antidevelopmental views of neo-Darwinian theorists are associated with philosophical and historical views that were articulated around the 1959 centennial of the publication of Darwin's *Origin*. These were not purely philosophical notions, of course. They involved central aspects of the population genetic understanding of evolution and the theory of heredity on which it was based. Nevertheless, I also trace the continued challenges from advocates of development through the century, and the alternative understanding of the nature of evolution on which those views were based.

The invention of SH is one central topic of Part II of the book. SH is also the ideological background against which Part I is set. The reader may want to read Part II first. It explains how SH came into being – that same SH that formed the problematic of Part I.

My identification of Synthesis Historiography as an actual trend of scholarship in history of biology may be controversial. The very fact that I give it a name suggests that I harbor a paranoid conspiracy theory. In fact, I don't believe in a conspiracy at all. The early examples of SH really were "in a good cause" as David Hull reports. In the 1960s, modern neo-Darwinism was still poorly understood by nonbiologists, and (it appears) especially by philosophers.[4] SH served the purpose of explaining the modern theory in a clear and concise way, revealing its philosophical richness, and expressing its opposition to theoretical alternatives. There is no doubt that the public and the academic community were educated by those writings. The subsequent growth of the fields of both history of biology and philosophy of biology can be largely credited to the efforts of SH authors. My own education in history and philosophy of biology owes a very great deal to these writings, as does everyone else's. Nevertheless, I believe that central features of those writings have outlived their usefulness. They are a barrier to a better understanding of the history of evolution theory, and possibly to an integration between the Synthesis and evo–devo. To take development seriously, we must challenge the historical interpretations that made it seem so irrelevant.

My defense against the accusation of paranoia can come only by the examples I discuss in Part I. The primary source of Synthesis Historiography is Ernst Mayr, one of the "architects" of the Evolutionary Synthesis and a giant in twentieth-century history and philosophy of biology. Other authors who have contributed to (what I see as) the Synthesis interpretation of history are philosopher David Hull and historians Peter Bowler and, to a lesser extent, William Coleman. Michael Ruse will hold a neo-Darwinian bias until the day he dies, but his historical writing poses somewhat fewer interpretive problems. I cannot identify SH with a definite list of authors, because some of the most egregious concepts have even been adopted by authors who are generally sympathetic to development. For example, both Dov Ospovat and Stephen Jay Gould follow Bowler in describing pre-Darwinian transcendental morphology as an "idealist version of the Argument from Design" (Bowler 1977). I consider this a serious misrepresentation, detrimental to structuralist interests,

---

[4] "The philosophers, in particular, were almost unanimously opposed [to Darwinian selectionism] until relatively recent years" (Mayr 1980a: 3). Mayr cites Ernst Cassirer, Marjorie Grene, and Karl Popper.

and I argue against it in Chapter 3. Nevertheless, Ospovat was the first historian to reveal the importance of pre-Darwinian structuralism, and Gould was among the strongest structuralist critics of the Synthesis (Ospovat 1981; Gould 2002).

The final reason I find it difficult to pin down SH precisely is that some of its effects are so global as to be virtually untraceable. Here is an example: Between Linnaeus and Darwin, most intellectuals believed in species fixism. What were their reasons? As Chapter 2 documents, species fixism was an empirically founded discovery of the mid-eighteenth century, not an ancient doctrine from Greek philosophy and Christian theology as SH would have it. Ernst Mayr acknowledges this fact in his early writing, although seldom after the crucial year of 1959. However, very little scholarly writing exists on this topic today (exceptions are cited in Chapter 2). This lack of scholarship is almost certainly due to SH itself. No one has looked for the reasons for species fixism *because they were already known*! The explanation given by Mayr in 1959 (mirrored by A. J. Cain in 1958, and echoed by Hull and Michael Ghiselin shortly after) was that pre-Darwinians were essentialists, and essentialism implies species fixism.

This is the Essentialism Story, the central pillar of SH (named following Winsor 2003). It holds that typology and essentialism (said to be identical doctrines) were to blame for the pre-Darwinian belief in species fixism, and later to blame for evolutionary saltationism and other theoretical alternatives to the Evolutionary Synthesis. The historical claim was not challenged until 1990, and it is still reported as historical fact (Ereshefsky 2001). We will see that it is both conceptually inadequate to account for species fixism and historically irrelevant to species fixism and other pre-Darwinian beliefs. Nevertheless, it seems to me, the reason that we have so little historical information about the reasons for the belief in species fixism is that the Essentialism Story was assumed to have answered that question. If this is correct, then SH is responsible for the *lack of historical research* on species fixism.

The Essentialism Story is so central to SH that I am almost willing to drop the SH concept itself and let my case stand or fall on the Essentialism Story. My reason for continuing to talk about SH is that the Essentialism Story *sounds like* a narrow issue. It is not. It is tied to substantive theoretical aspects of the Evolutionary Synthesis, to the Synthesis reconstruction of the history of evolution theory, and to Synthesis critiques of modern structuralist, developmentalist alternatives. Other aspects of historiography, like the treatment of transcendental morphology and the alleged "idealist version of the Argument from Design" also fall under the heading of Synthesis Historiography. Therefore, I will continue to allude to Synthesis Historiography.

13

Three issues of general epistemological orientation recur in this discussion. The book is not intended to justify one or another philosophical position; the philosophical issues are secondary to those of scientific method and content. However, philosophy played a role in the debates, and these concepts were (and are) involved. Two of the concepts roughly correlate with the traditional empiricism–rationalism dichotomy. They are what I will call *inductivist caution*, and *idealism*. The third is essentialism itself, and the two distinct versions of essentialism that have been in play during the past half-century.

1.5.1   Inductivist Caution

This view is the often-empiricist conservative methodological position that encourages the scientist to avoid speculation, and especially to avoid *conceptual* speculation, speculation that invents new concepts. The institutions of science seem to vacillate between extremes on the conservative–liberal spectrum. Larry Laudan contrasts "epochs when the object of science is seen as discovering empirical laws" and "epochs when the stress is upon discovering explanatory, deep-structural theories" (Laudan 1980: 178–179). Laudan says that the early twentieth century was conservative, with logical positivism and the Copenhagen interpretation of quantum mechanics (I would add behaviorist psychology). The late twentieth century is far more liberal, with quarks and black holes (and cognitive psychology with dozens of hypothesized mental modules). David Hull has documented the conservative and empiricist atmosphere of Britain in the pre-Darwinian period (Hull 1973, 1983; one of many debts I owe to SH authors).

Inductivist caution is the attitude that values high-probability inferences that remain close to the scientist's direct observations. *Phenomenal laws*, those that are expressed in terms of observable properties, should be the ordinary scientist's goal. Once an adequate understanding of the world in terms of phenomenal laws was available, one might try to discern an underlying *causal law*, which designates a true cause or *vera causa* (Ruse 1979). The distinction between phenomenal laws and causal laws seems always to have been illustrated with Kepler and Newton. Kepler discovered the phenomenal laws that describe the motions of planets; it took a Newton to determine the vera causa of the law of universal gravitation (Hopkins 1860). Even though causal laws were sometimes reachable, it was considered immodest for an ordinary scientist to claim to have reached one. This view of vera causa was certainly an influence on the severe reaction that most British scientists had to the anonymous publication in 1844 of *Vestiges of the Natural History of Creation*, and it had similar affects on reactions to Darwin's *Origin* in 1859. Darwin's claim of common descent was much less objectionable than his claim to have found its cause in natural selection.

The empiricism behind inductivist caution will play a role in other contexts than that leading up to Darwin. The view is often associated with an antirealistic attitude toward unobserved theoretical entities or processes. As Laudan reminds us, we live in a time of lively interest in such inferences. However, conservative eras are frequent in the history of science, when the payoffs of high speculation are believed to be outweighed by its risks. Conservatives insist that scientists stick to observations and predictions, and abandon hypotheses and explanations. Although this kind of operationalism or positivism is out of fashion today, it is historically important. It can smooth the transition between old established concepts and radical new ones. This is not an intentional strategy of scientists in these cases; their positivism is an honestly held methodological principle. Nevertheless, the positivism of the transitional scientist often gives way to realism about the radical new theory. A declaration that a bizarre new idea is "just a calculating device, not a description of reality" can calm the resistance against a new idea long enough for the progress of science to work its way past the barrier. Newton's "hypotheses non fingo" was his way of dodging the fact that gravity was action at a distance, believed to be metaphysically impossible in his day. Copernicus's editor Osiander inserted a preface to *De Revolutionibus* that said the sun-centered system was not about a real earth moving through space around a real sun; it was a mere calculating device. After the shock had worn off, the notions of a gravitational force and a moving earth were interpreted realistically. In the episodes to follow, we will see that the origins of both embryology and genetics were aided by similarly conservative, antirealistic stances that buffered the objections that otherwise might have been fatal to radical new theories.

One complication must be introduced into the discussion of empiricism and inductivist caution. In periods during which new theories are being developed, scientists are often willing to commit themselves to a phenomenal law, and to the claim that the law points to an important underlying causal explanation, but they are not ready to commit themselves to the nature of the underlying explanation itself. I call this position *cautious realism*. Cautious realists believe in a reality underlying a phenomenal law, but they are not yet ready to name it. This was surely the position of Kepler with respect to his planetary laws. I will argue that it was also the position of many pre-Darwinian naturalists with respect to the organic patterns they were discovering. The SH tradition tends to label pre-Darwinian authors as antievolutionists (essentialists, special creationists, etc.) when they merely *fail to assert* evolution as the cause of the regularities that they studied. I will argue that many of these individuals were instead cautious realists. When we recognize cautious realism as a respectable scientific stance, we need no longer divide the nineteenth century into evolutionists and creationist–species fixists.

1.5.2   Idealism

An important theoretical orientation in early-nineteenth-century biology is variously labeled *idealistic morphology, transcendental anatomy*, and sometimes *philosophical anatomy* (Rehbock 1985). Like many philosophical terms, "idealism" can mean many different things. In this case, it refers to the Kantian variety. The topic is treated with true derision by many authors in the SH tradition.[5] One SH approach that is *not* helpful is to define 'idealism' in terms of essentialism or typology. For one thing, the assertion is simply false. For another, the Essentialism Story attributes essentialism to almost all pre-Darwinians, so it doesn't help us to understand idealists when we hear that they are essentialists. Finally, it is very difficult to explain the philosophical context of Kant and his relation to the continental morphologists without becoming embroiled in arcane vocabulary and disputes that are far from our topic.

I am frankly going to duck this issue. I will discuss the philosophical underpinnings of the continental morphologists only enough to allow their scientific views to be presented with minimum prejudice. In my view, the scientific work has value enough to incline the reader toward tolerance for the metaphysical idiosyncrasies of its authors. For a deeper understanding of the philosophical complexities of the movement, I must refer the reader to other sources (Lenoir 1982; Sloan 1992, 2002; Larson 1994; Asma 1996; Richards 2002). Now a few comments to comfort the squeamish reader, who will soon be asked to take transcendental anatomy and idealistic morphology seriously.

---

[5] Brief allusions include "the lofty fallacies of idealist philosophy" (Mayr 1976: 258) and "idealist moonshine" (Bowler 1984: 125). Slightly more useful is the description of idealists as those who "tended to explain the order in nature by reference to ideal types" (Hull 1973: 67). The most extended discussion is by Bowler, who associates idealism with Hegelian political theory (subordinating the individual to the state) and traces this view through to Karl Marx (Bowler 1984: 99–102). Bowler cites Karl Popper on his Hegel commentary; Popper is also the source of the "essentialism" epithet that will loom large in my discussion. The Popperian origins of the notion of essentialism suggest that the Essentialism Story itself is a holdover from the epistemologically conservative first half of the twentieth century.

The extreme SH antagonism toward Kantian idealism seems to be a holdover from the epistemologically conservative first half of the twentieth century. Both Newton and Kepler had far zanier metaphysical ideas than the Kantian idealists, but their scientific reputations are unsullied. The Essentialism Story was introduced around 1959, just as logical positivism was drawing its last breath. It has kept alive an animosity toward biological idealism that is unlike anything seen in other areas of the history of science. This is especially striking when one recognizes how very different, and very Kantian, both the philosophy of science and cognitive psychology had become by the 1970s. Condemnation of the idealists was easy for the positivists but much harder for a post-Kuhnian philosopher of science. Kant held the mind to be active, producing "ampliative" judgments that went well beyond the input of the senses. What modern philosopher of science could disagree? Modern psychology is even more Kantian than philosophy of science. Several prominent commentators have argued that cognitive psychology, in describing the mind as active and constructive, has empirically confirmed the basic framework proposed by Kant (Guyer 1987; Kitcher 1990; Brook 1994).

As an aid to the reader, I therefore propose a version of idealism intended only to open his or her mind to the morphological theories called "idealist." It is Amundson's Minimal Idealism.

1. Human epistemic abilities are not limited to inductively generalizing from sensory experiences; they are able to create hypothetical constructions of unobserved reality.
2. Let us call these hypothetical constructions *ideas.*
3. Once constructed, these ideas can be compared with sensory experience, which may confirm or deny the value of the ideas.

On this account, ideas are nothing but hypotheses (although they are unbound by empiricist restrictions on their content). There is no requirement for ideas to have a distinct ontological mode of existence, in God's mind or anywhere else. Ideas are merely what we (i.e., our minds) use in understanding the world. This is almost all that we need for so-called idealist morphology to make sense. I am skipping many details, of course. Some of Kant's ideas were automatic instincts of the mind that themselves *formed* (rather than following from) experience. Idealists differed on the status of an idea once it had been constructed, and there were great differences about which ideas provided (or constituted) knowledge, and which ideas were only heuristic, regulative principles that might be indispensable for research. Kant held that teleology, for example, was an indispensable but "regulative" (heuristic) assumption, needed for the practice of biological research. Others adopted what Sloan calls "the Schelling revision," and declared that Kantian ideas like teleology were "constitutive" of nature (Sloan 1992: 33). These were the *Naturphilosophen*. They have received immense abuse through history for their metaphysical merriments, beginning as early as the 1820s. However, even this distinction will not concern us. We will be concerned with the value of the scientific concepts that Naturphilosophen and other idealists contributed to biological thought. We will fret over their metaphysics no more than we fret over Kepler's.

Amundson's Minimal Idealism distorts the history of philosophy. Nevertheless I hope it allows the reader who might be inclined to dismiss idealist morphological theories to pay them heed. SH authors are not concerned with the purely philosophical aspects of idealism anyhow. They oppose idealism because it leads (they say) to faulty scientific theories. I disagree. Whatever the value of idealism as a philosophical doctrine, it has given rise to extremely important scientific theories. These theories played a crucial role in the history of evolutionary thought, and they do not deserve the disdain to which they have so long been subject.

1.5.3   Two Essentialisms

Essentialism is a doctrine about natural kinds. Natural kinds are assumed to possess essences. The nature of the commitment to essences differs in the traditional concept of essentialism (embedded in the Essentialism Story) and a modern version discussed in the paragraphs that follow. In its traditional form, essences are definitional sets of intrinsic necessary and sufficient conditions that logically determine an entity's membership in the kind. Kinds are (by their definitional essences) discrete from one another. Because the definitions are timeless, the kinds are eternal and unchanging. The essential characters of some kinds (e.g., biological taxa) may require scientific discovery. However, essentialism itself is not a scientific doctrine but an a priori doctrine of metaphysics. It is not an empirical but a conceptual truth that natural kinds exist, and that they are fixed, eternal, and changeless.

Since about 1970, philosophers of science (though usually not philosophers of biology) have discussed essentialism from a very different perspective. Philosophers have increasingly been attracted to scientific realism, consistent with Laudan's observation about our epistemological era. Realism has become associated with a kind of essentialism that is quite distinct from that of the Essentialism Story. Natural kinds are seen as categories that play a role in a law of nature (after Quine 1969). Both the determination of the kinds themselves and the discovery of the causal facts underlying their kindhood (their essences) are seen as a posteriori empirical achievements. Essences are conceived not as definitional stipulations but as underlying causal structures that explain the observed, discovered lawlike facts about the kind. The standard example of a natural kind is a chemical element such as gold. The essence of a chemical element is the atomic structure of an atom of that element. Richard Boyd has prominently argued for a version of this kind of essentialism (Boyd 1991). He distinguishes two different kinds of realist–essentialist definitions. One definition is suitable to the early stages of the development of a scientific theory. This is a "programmatic" definition, which specifies the role that a particular natural kind is intended to play in a theory. An example is the position on the periodic table of a chemical element, indicative of its combinatorial relations with other elements. The other definition is "explanatory," and it typically occurs later in the development of a science. It specifies the essence as the causal properties in virtue of which the kind is able to fulfill its programmatic role in the scientific theory. The programmatic definition of chemical elements involves their valences and combinatorial relations. The explanatory definition specifies the atomic structure that explains the valence and combination relations. Chemical elements show the combinatorial relations that they do in virtue of their different atomic structures.

Boyd's version of essentialism is quite distinct from that of the Essentialism Story. He carefully explains the difference: The Essentialism Story describes a metaphysical, aprioristic account that implies fixed and unchangeable kinds and is based on strict necessary and sufficient definitions. Boyd's version is thoroughly a posteriori. It freely uses fuzzy definitions (which correspond to "homeostatic property clusters" rather than strict necessary and sufficient conditions) and has no implications regarding fixity. In the Essentialism Story, essences are not offered as causally explanatory. Essentialism is treated instead as a pre-existing, a priori, rationally unacceptable metaphysical commitment that entailed species fixism in the entire absence of empirical evidence.

I introduce Boyd's version of essentialism not because it is relevant to the Essentialism Story. As Boyd insists, it is very different. Boyd does not challenge the story (as I will); instead he merely claims that *those* essentialists were doing something different than he is doing (Boyd 1999: 146). I introduce Boyd's version because it expresses a form of scientific realism that will play a genuine role in the discussions that follow. Scientific talk about essences need not imply the Essentialism Story (metaphysical, semantic, a prioristic, and species fixist). It may instead indicate a commitment to scientific realism regarding a process or entity, even when the deep or ultimate nature of the entity is not yet understood. I will argue that the Essentialism

Story is false. However, we will see in nineteenth-century biology some signs of a Boydian essentialism, and the use of programmatic definitions of kinds. This fits well with what I have already called cautious realism. Realistic but a posteriori commitments to certain entities (including both species and the Natural System itself) were an important and progressive feature of pre-Darwinian thought. SH makes it impossible to recognize this important scientific development.

1.6 EXPLANATORY RELATIVITY

Scientific explanation is an epistemological topic of a different nature than the relation between theory and evidence. Explanations are context relative in a way that theories are not. They are best understood as answers to why-questions (Bromberger 1966). If a person asks you "Why *x*?," you may respond not with an immediate answer but with another question: "What is it about *x* that you don't understand?" An explanation of *x* that is satisfactory to one person might not be satisfactory to someone else, because "what they didn't understand about *x*" was different. Sometimes these differences can be discovered by rephrasings of the why-question. "Why did Adam eat the apple?" might mean

> Why was it Adam who ate the apple (rather than someone else)?
> Why was it the apple Adam ate (rather than some other food)?
> Why did Adam eat the apple (rather than throwing it, or cooking it into a pie)?
> (after van Fraassen 1980)

The Adam example illustrates the fact that explanations occur against a background of presuppositions that can be seen as a "contrast class" of alternative possible answers assumed by the questioner. The questions {Why *x* rather than *a*, *b*, or *c*} and {Why *x* rather than *d*, *e*, or *f*} call for different answers. Another illustration comes from the apocryphal story about the notorious bank robber Willie Sutton. A priest visited Willie in his cell and asked him why he robbed banks. Willie replied, "Because that's where the money is." The priest assumed the contrast class of robbing banks versus making an honest living. Willie assumed the contrast class of robbing banks versus robbing other establishments.

The interest-relativity of explanations has led many philosophers, including Bas van Fraassen, to the view that explanation is a matter of practical or applied science, not of "pure science." Van Fraassen ends his elegant discussion of explanatory relativity by claiming that a successful explanation gives no special support to the scientific theory that underwrote that explanation (van Fraassen 1980: 156). Evidence that the theory *is true* must come independent of the fact that it *satisfies our interests*. I have no desire to argue that point: What is and is not pure science is not my concern. Instead, I am interested in understanding the conflicts between contrasting scientific traditions. Here, explanation is crucial.

If you want to understand a scientific theory or research tradition, look at what the scientists want to explain, and what they think counts as an explanation of it. If you want to understand the difference between a pair of competing scientific theories, look at why each theory's advocates believe that the others' explanations are faulty.

Scientific theory changes are often associated with changes in what is thought to require explanation. Aristotle said that *motion* required explanation; Newton said that *change of motion* required explanation. Phlogiston theory attempted to explain the observable qualitative characteristics of chemical compounds in terms of the qualities of their elements: Metals are shiny and malleable because they contain phlogiston, the element of fire. Atomic theory did not replace the phlogistic explanation of the qualities of compounds. Instead, it abandoned the goal of explaining the qualities of compounds entirely, and it replaced it with the new goal of explaining the proportionate weights of elements and compounds – a poor substitute, to my mind. Nevertheless, the new explanatory goal was achievable, and the old goal was not. To understand the differences between these theories, we must understand the differences in the questions they asked – which is to say, the explananses for which they sought explananda.

The methodological differences between research programs are often hidden behind the verbal similarities in the questions they ask. This can lead to apparent conflicts where no conflict really exists. (Imagine a debate between priests: "Willie robs banks because that's where the money is!" "No, he robs banks because he has no moral conscience!") It is important to be aware of these potential confusions. Aristotle's four causes are really different kinds of explanation; each may apply to the "same" explanans (or rather to four different explananses that have the same verbal expression). Ernst Mayr's distinction between proximate and ultimate causation (discussed in detail in Chapter 10) is similar. Proximate and ultimate explanations need not conflict. The choice between them depends on one's explanatory interests.

Explanatory relativity is important for this project because the dichotomy of function versus structure is a contrast of explanatory modes. It is possible to have both a functional and a structural explanation of the "same" phenomenon. Nevertheless, functionalist and structuralists tend to clash, and the clash runs through our entire 200-year historical narrative. Function versus structure is a contrast not simply about the facts of the world but about what phenomena require explanation. The advocates of one side will frequently announce that their opponents "do not give explanations, but only descriptions." This is a sure sign of explanatory relativity. A large part of my task will be to tease apart these disputes and try to separate those that involve actual matters of fact from those that are based on disagreements about *what needs to be explained.*

Explanatory relativity does not imply a relativity of truth, or even a relativity of theoretical adequacy. Some theories really are better and some are worse. Structuralist theories are better in some domains, and functionalist theories in others (Amundson 1989). It is difficult to understand the contrast until the differences in explanatory goals are understood. The differences in explanatory goals between neo-Darwinism and evo–devo have not been adequately recognized. I will develop an interpretation of these differences that is based on their respective historical origins, but I must revise the histories before such a comparison will be useful.

1.7   HISTORICAL CONVENTIONS

Professional historians of science will soon recognize that a philosopher wrote this book. It sometimes reads as if history is made up of the interactions of ideas rather than of people. It will sometimes overgeneralize the similarities and underestimate the complexities of historical events and debates. This the hazard of doing history and philosophy at the same time. I have had to ignore some particulars in order to recognize the general and repeated aspects of methodological debates through history. I have tried not to use this as an excuse to misrepresent the debates, but it has forced me to neglect some complexities.

I have also, for convenience's sake, used some terms in ways that are historically inaccurate. Here I apply the term *Lamarckian* to the theory of the inheritance of acquired characteristics, even though there is very little historical justification for that label. (Lamarck didn't invent the notion, and prior to August Weismann almost everyone including Darwin believed in it anyhow.) A second misused term is *neo-Darwinism.* This was first applied to Weismann's evolutionary theory, which accepted natural selection and rejected Lamarckian (ahem) inheritance. Some historians argue that neo-Darwinism should not be used to label twentieth-century evolutionary theory (Reif, Junker, and Hossfeld 2000). Notwithstanding, here I use it to label the theory associated with the Evolutionary Synthesis. Finally, historians are careful not to apply invented terminology to historical cases that existed prior to the invention. So, for example, Newton would not be called a scientist because William Whewell invented the term *scientist* in the 1830s. I violate this convention with impunity, but I try to avoid the errors that the convention is designed to prevent.

1.8 HISTORICAL PRÉCIS

A central theme of the book is structure versus function. I once had hopes of reducing two centuries of history of biology to this theme (a typical philosopher's ambition), but it didn't work. Two additional factors make the story much more complex. One is a second difference in explanatory goals (besides function–structure) between Darwinian and structuralist evolutionary theories. The other is a radical change in the concept of heredity that occurred in the early twentieth century.

Even aside from their commitment to functionalism (adaptationism), Darwin and the Evolutionary Synthesis share an explanatory goal that separates each from structuralist theories. I begin to examine this contrast in Chapter 4 and continue throughout the book: Darwinian theories are *change theoretic* whereas structuralist theories are *form theoretic*. Very roughly, this implies that Darwinian theories do not accept responsibility to explain form (or anything else) in the evolutionary entity; rather they explain change in the entity through evolutionary time (change in form, or change in anything else). The ancestral features are assumed as a background condition. In contrast, structuralist theories accept responsibility to explain form both in ontogenetic and phylogenetic time, and to associate the two aspects of form-explanations. So ancestral form is not merely a background condition for structural theories; it must receive an appropriate (e.g., ontogenetic) explanation. The change-theoretic explanatory goal is shared by Darwin and the Evolutionary Synthesis. This legitimates the traditional claim that the Synthesis is Darwinian (more than the questionable claim that Darwin was a population thinker).

The second factor needed to understand the twentieth-century history of the relation between evolution and development is a certain radical change in the concept of heredity. I literally remember the hour I came to realize this fact. It was in 1997 at the Seattle meetings of the ISHPSSB, at a session honoring historian Fred Churchill.[6] During the discussion, one of Churchill's

---

[6] ISHPSSB is the International Society for the History, Philosophy, and Social Studies of Biology, pronounced "Ishkabibble."

former students said something to this effect: "As Fred has shown, during the nineteenth century heredity was considered to be an aspect of embryological development." My head swam. I quickly chased down Jane Maienschein to get the proper citation (Churchill 1974). *This* was how a structuralist view of evolution could seem almost self-evident in the nineteenth century but be labeled as incoherent in the twentieth. (I had already rejected the Essentialism Story that blamed structuralist beliefs on silly metaphysical views.) A related but independent historical discovery was made by Jean Gayon (Gayon 1998). Gayon showed that natural selection in the nineteenth century was inconsistent with any known theory of heredity. Moreover, no one even knew how to specify what must be true about heredity in order to make natural selection work as an evolutionary mechanism! So the nineteenth-century rejection of natural selection as the primary mechanism of evolution was based not on typological dogma but on hard-nosed experimental science (and the Essentialism Story fails again). Thomas Hunt Morgan introduced the new kind of heredity that distinguishes it from development (Morgan et al. 1915). This new entity (nondevelopmental heredity) would turn out to enable natural selection as a long-term evolutionary mechanism but at the same time disallow the relevance of development to evolutionary change. This story and the consequent debates comprise Part II of this book.

I intend the overall narrative of the book to come out right for evo–devo. Evolutionary developmental biology has a respectable history of intellectually productive predecessors from the early nineteenth century through the twentieth. Unfortunately, so does neo-Darwinism. I feel that I have successfully refuted the SH belittlement of nineteenth-century structuralist evolution theory. I have defended twentieth-century structuralists against their methodological critics. I have even shown some interesting details about methodological contrasts between neo-Darwinism and evo–devo. But I have not shown neo-Darwinism to be wrong about evolution, and I have not shown how to meld the two traditions into one. That must remain a job for the scientists.

Here is a sketch of the historical narrative to follow.

PART I: DARWIN'S CENTURY: BEYOND THE ESSENTIALISM STORY

1. Species fixism was a discovery of the eighteenth century. It was scientifically progressive because it enabled the construction of a classificatory Natural System. A Natural System would have been incoherent under prefixist understandings of organic relationships.

2. Contrary to SH, essentialism and typology were neither part of the grounds for species fixism nor involved in the early systematic attempts to classify organisms into the Natural System.
3. Morphology (including embryology) grew out of systematics, but it developed into an autonomous discipline. It originated the important explanatory goal of the explanation of form. The concept of morphological types grew from this enterprise. Morphological types are not appropriately characterized by the SH condemnation of typology for reasons to be discussed.
4. Early understandings of the Natural System were conventionalist in the sense that taxonomic categories were seen not to reflect objective reality but instead to organize data for human uses. As time passed, the Natural System came more and more to be seen as an accurate depiction of real relationships in the natural world. Morphology was centrally involved in this increasing *reification* of the Natural System.
5. Darwin's achievement must be seen in two parts. Part 1 (by his own classification) was the argument for natural selection as the force behind evolutionary change. Part 2 was his proof of descent from common ancestry. Part 2 was successful almost immediately. The general acceptance of Part 1 awaited the Evolutionary Synthesis seventy years later.
6. Darwin's proof of Part 2 depended on the existence of a well-established and reified Natural System. It also relied heavily on the morphologists (including their brand of typology), and Darwin acknowledged that fact. SH authors assert that Darwin rejected the importance of, for example, Richard Owen's typological morphology. These reports stem from systematic misreadings of Owen and of Darwin.
7. Darwin's argument for Part 1 introduced a new explanatory goal for evolutionary biology. It is the explanation of change. Darwin did not accept the morphologists' (e.g., Owen's) goal of explanation of form (though the distinction was not well recognized by either side). In changing the explanatory goal, Darwin abandoned the explanatory aspects of the morphologists' concept of types. This innovation (more than population thinking) justifies viewing Darwin as forefather of the Evolutionary Synthesis.
8. Evolutionary morphology was the first evolutionary research tradition. It accepted Darwin's Part 2 but marginalized Part 1. It incorporated common descent into the traditional morphological goal of the explanation of form. The program failed but was not conceptually flawed by typology in the way that SH commentators report it to be.

PART II: NEO-DARWIN'S CENTURY: EXPLAINING THE ABSENCE AND
THE REAPPEARANCE OF DEVELOPMENT IN
EVOLUTIONARY THOUGHT

9. The concept of heredity grew in importance during the nineteenth century, largely because of interest in Darwin's theory of natural selection. Heredity was conceived as an aspect of embryological development. This is very different from the modern conception of heredity, and the significance of that difference is enormous.

10. Natural selection remained scientifically anomalous throughout the nineteenth century. This was (not because of the bad influences of typology but) because the known facts of heredity could not be shown to be consistent with natural selection as a long-term cause of evolutionary change.

11. The separation of heredity from development was achieved around 1915 by T. H. Morgan and associates. It required a quasi-positivistic argument that hereditary properties of the germ could be said to "cause" adult traits *at a distance* (i.e., ignoring the intervening causes that construct the adult from the zygote). Even though we all know that ontogenetic processes are involved in the construction of adult traits, those processes are henceforth to be ignored when we are discussing heredity.

12. Most embryologists rejected Morgan's quasi-positivistic gene on the grounds that it cannot, in principle, explain development. In effect, they rejected the new antidevelopmental definition of heredity.

13. In contrast, population genetics was based entirely on Morgan's positivistic "transmission" gene. For the first time, a theory of heredity was consistent with natural selection as a cause of long-term evolutionary change. (This required changing the meaning of *heredity* of course.) Thus, Morgan's quasi-positivistic gene *enables* natural selection as an evolutionary cause and at the same time *disables* the participation of embryology in the resulting evolutionary theory.

14. The Evolutionary Synthesis is formed on the basis of population genetics. It shares Darwin's goal of explaining change. Meanwhile, experimental embryology flourishes while largely ignoring genetics (including the nonpositivistic "developmental genetics"). It retains the morphological goal of explaining form. A few individuals (such as Goldschmidt, Waddington, and Schmalhausen) attempt to integrate the explanation of form into the Synthesis, but they fail.

15. By the 1950s, support for the molecular genetic studies of development is growing. This has two independent effects. The first effect is that nongenetic experimental embryology fades.

16. The second effect of the rise of molecular genetics is that, around 1959, Ernst Mayr, leader of the naturalist contingent within the Evolutionary Synthesis, is moved to write several methodological papers on the importance of naturalistic studies in evolution. These are aimed not particularly at refuting the relevance of development, but at enhancing the status of naturalists in the Synthesis (as against mathematical geneticists), defending the Synthesis against critics, and aligning it with Darwin as an intellectual forefather. This is the origin of SH. (Needless to repeat, SH misreports several of the historical episodes listed herein.)

17. Mayr in 1974 organizes two symposia on the history of the Evolutionary Synthesis (reported in Mayr and Provine 1980). During the 1970s and 1980s a series of attacks are mounted against the Synthesis, alleging methodological flaws. One of these is the structuralist claim that the Synthesis ignores development. The Synthesis is writing its own history just at the time that it is being attacked.

18. In the 1980s and 1990s, Synthesis advocates deploy a series of arguments to defend against the continuing structuralist criticisms. These include concepts derived from Morgan's 1915 distinction between heredity and development, and others from Mayr's important methodological writings around 1959 (Mayr 1959b, 1959c, 1961). It is argued that development is conceptually irrelevant to evolution. These arguments can be seen to revolve around the difference in explanatory goals between explaining change and explaining form. For this reason they are inconclusive.

19. Tremendous growth of molecular genetics in the 1990s provides much new developmental data. These data show far more unity within the embryological creation of organic form than the boldest structuralists had expected. Evo–devo is generated. Nevertheless, the explanatory contrasts with Synthesis evolutionary theory remain.

20. Amundson begins to write his revisionist history of evolutionary theory, offering an alternative to SH and attempting to understand the methodological conflicts during the twentieth century between evolutionary and developmental biology. . . .

I

Darwin's Century

Beyond the Essentialism Story

2

Systematics and the Birth of the Natural System

This chapter sketches the development of the practice of systematics and taxonomy from the time of Linnaeus through the time of Darwin. The central
topic is the origin and growth of the Natural System. Development of the
concept of the Natural System during that period was crucial to Darwin's enterprise. My narrative differs from many others in the ways I see the Natural
System to have changed. Traditional reports of pre-Darwinian systematics
claim that belief in the fixity of species is ancient. It is said to be founded
on the ancient Greek metaphysical principles of essentialism, typology, or
both. These principles asserted that natural kinds (including species, higher
taxa, chemical elements, and even geometric shapes) were characterized by
essences that were distinct and unbridgeable. Biological species had fixed,
defining characters in the same sense that geometric figures did. Just as triangles must have three sides, species members must have the characters essential
to their species. Just as squares and triangles are separated by an unbridgeable
gap reflected in their essential characters, so are species. Darwin's achievement was to overthrow the essentialism–typology doctrine that had governed
earlier taxonomic practice, to recognize the variability within species and the
continuity between them. This allowed evolutionary transitions to occur. This
attribution of species fixism to ancient Greek metaphysics will be termed the
Essentialism Story. The Essentialism Story is not just a report of a historical
belief but also an *explanation* of another belief. The pre-Darwinian belief in
species fixism is *explained by* the belief in essentialism–typology.[7]

---

[7] Sometimes essentialism is used as a mere synonym for species fixism. This is not the Essentialism Story. The founders of the Essentialism Story applied it primarily to the species level.
Nevertheless, they asserted that it applied equally to higher taxa and to other kinds such as gold.

31

In contrast to traditional histories, I argue for the following: (1) species fixism was not an ancient doctrine and was barely a century old when it was refuted by Darwin; (2) fixism was scientifically progressive and a necessary precursor to Darwin's achievement; (3) the Essentialism Story is false of systematic practice between Linnaeus and Darwin and was never an important ground for species fixism; (4) although typology takes many forms, it was almost never in direct opposition to Darwin's evolutionary thought; (5) in one of its forms, typology was crucially progressive toward a belief in evolution.

The adoption of species fixism in the mid-eighteenth century enabled the beginnings of the concept of the Natural System. In its early versions, the Natural System itself did not lend itself to a Darwin-style evolutionary interpretation. A century of development was needed before the Natural System took on a form that Darwin could present to the world as a depiction of the genealogical relations among species. We can see in retrospect how the theoretical debates of the early nineteenth century prepared the way for Darwin. However, these debates were not about evolution itself, and especially not about the variability and modifiability of individual species. Instead, they were about the nature and proper interpretation of the Natural System. Was the system seen as a human contrivance, intended to capture and arrange a large body of individual facts about species? Did it, instead, represent the discovery of real relationships and discontinuities within the living world itself? These two ways of seeing the Natural System reflect the contrast between nominalism (or instrumentalism, or conventionalism) and realism in the philosophy of science. Nominalists regard a good theory as an economical summary of data and as an instrument for predicting future data. Realists regard a good theory as an ostensibly correct representation of objective reality – a reality that lies beyond the data themselves.

The nominalism–realism contrast is far too simple to reflect the complexity of the various views on the Natural System, but it does provide a framework. Nominalists about the Natural System were not philosophical antirealists; they were nominalists only in a restricted sense. I refer to them as *taxonomic nominalists* to reflect the fact that the taxonomic categories in the Natural System were not considered to represent objectively real relationships. Almost all of them had important nonnominalist commitments to the nature of species, the history of life on earth, and the proper explanation of the patterns of diversity that were observed and recorded in the Natural System. Taxonomic nominalism amounted to the denial that the Natural System revealed

The wide popularity of the story is surely due to its generality (see the discussion in Sober 2000: 148 ff.).

real relationships between species in nature; taxonomic nominalists denied that taxa are *real groups*. Most were realists about species – species exist as distinct and unique entities in the world. Many believed that similarities in the forms of species are properly (objectively, realistically) explained by similarities in function. However, the similarities that make the construction of the Natural System possible are *mere* similarities, coincidences that are due to function. They are not indicators of any sort of real relationship beyond the coincidences of function. One author expressed taxonomic nominalism this way: Each species exists *per se*; it does not exist *as* a member of a genus, family, or class. To a nominalist, the Natural System is natural only in the sense that it reflects the pattern of similarities that are actually observed in nature. Taxonomic categories are created for human consumption. They do not reflect real relatedness in nature.

During the first half of the nineteenth century, a number of factors began to encourage realism about the Natural System. Many naturalists came to believe that classification revealed real groups of species, not merely coincidences of similarity. Many grounds for this increasing realism are discussed in the following three chapters. However, the most important fact about the realists is a negative fact: The emerging belief in real groups was not associated with any particular commitment to an "ultimate cause" for species' relationships within the groups, or to an ontological status of the groups themselves. The ultimate nature of the natural groups was not a settled matter.

This illustrates a second aspect in which I believe traditional reports on pre-Darwinian systematics are faulty. The traditional narrative assumes that naturalists who explicitly assert the *reality* of an entity (e.g., a type) must make a metaphysical commitment to some ultimate theory that explains and constitutes that reality. For this reason, it has often been assumed that pre-Darwinians who assert the reality of types must have been committed to the view that these types exist in God's mind. This is historically false. The dominant view among those who believed in *real groups* was that *we do not yet know wherein their reality lies*. These people were realists, but realists about what? They held the view that I call cautious realism. They were committed to the reality of a kind of thing (a natural group, sometimes a taxon, sometimes a type) even though they did not pretend to understand its deeper nature. In fact, this cautious realism is extremely common in the history of science. Galileo was realistic about planetary orbits even though he did not know what caused or maintained them. Mendeleev was realistic about the periodicity of the elements even though he did not know what caused that periodicity. Why should we deny to the believers in pre-Darwinian types that perfectly reasonable caution? The answer, I think, is that the traditional narrative of pre-Darwinian systematics comes from SH, a research tradition that is biased against certain important kinds of pre-Darwinian thinking. The traditional narrative overlooks the progressive aspects of species fixism, sees essentialism and typology where they do not exist, and therefore fails to recognize the importance of the realistic reinterpretation of the Natural System for Darwin's revolution.

Some of the analytical categories used in this chapter are sufficiently novel that I will specify them here at the start.

- Taxonomic realism is the view that classification schemes pick out real groups in nature; these are taxonomic entities (genera, classes, families, sometimes types) that are interpreted as objective entities. The species within a real group have real relationships with other species in that group that they do not share with species outside of the group.
- Taxonomic nominalism is the doctrine that real groups of species do not exist. Each species exists per se, in and of itself. The shared similarities on which taxonomy is based are coincidental, and taxonomic systems themselves are to be interpreted instrumentally or conventionally.
- Cautious realism with respect to a hypothetical entity is a commitment to the reality of the entity itself, without any deeper commitment to its ontological nature or its ultimate origin. I show that the naturalists who believed in real groups took a cautiously realistic stance in doing so.
- The Essentialism Story is the widely held historical view that the belief in species fixism among pre-Darwinian naturalists was due to their commitment to the ancient Greek metaphysical doctrines of Aristotelian essentialism or Platonic typology. I examine the historical grounding of this doctrine and find it wanting.

2.2   THE DISCOVERY OF SPECIES FIXISM

Modern narratives of the history of evolutionary biology take place against the background of species fixism. The story goes like this: The historical discovery of evolution was the overthrow of species fixism. From ancient days, Western intellectuals had conceived of a stable and unchanging world that had been created by God in pretty much the condition it now exists. Beginning in the early seventeenth century, traditional beliefs were shaken by a series of challenges to the world's constancy and stability. First the earth lost its stable location when astronomers sent it spinning through the heavens. Next the earth lost its stable shape, as continents and oceans were revealed as merely the latest

34

stage of a churning geological history. Nevertheless, in the early nineteenth century, life itself was still somewhat trustworthy. The ancient stability of the world could still be seen, if nowhere else, in the constancy of species. In this narrative, the fixity of species was the last vestige of the stable and unchanging world of the ancients. Christian theology had derived the fixity of species from the Genesis account of creation. Fixity had been underpinned by the doctrines of Platonic idealism and Aristotelian essentialism. The rediscovered ancient Greek texts dominated European thought from the Middle Ages onward, and they reinforced the biblical version of fixism. Variation was sometimes recognized within a species of organisms, but variation was believed to be strictly limited and never a threat to the stability of species. The fixity of species was the biological equivalent of the fixed earth in the center of the universe. Darwin's job was like that of Copernicus – the overthrow of an ancient belief in stability.

That's the story, but it's not true. The Western tradition was indeed centered on an unchanging world, but the fixity of species was not a part of that world. It may come as a surprise to the reader (as it certainly did to the author) that the fixity of biological species is not an ancient belief. It became widely accepted for the first time *both among naturalists and theologians* during the eighteenth century, only about a century before Darwin (Zirkle 1951: 48–49; Zirkle 1959: 642). Carl Linnaeus is widely known for his unequivocal statements of species fixism and special creationism. It is less widely recognized that Linnaeus was one of the innovators of fixism. Prior to Linnaeus and his botanical colleagues, beliefs in transmutation and spontaneous generation were extremely widespread.

This does not imply that earlier thinkers were evolutionists in anything like the modern sense. Species fixism and evolutionism are only two of many ways of conceptualizing the relations among different kinds of organisms. Prior to the establishment of species fixism, naturalists, theologians, and common people held a dazzling variety of transmutationist beliefs. The popularity of early transmutationism is so surprising to most modern readers (including the author) that it is worthwhile examining some of these old beliefs.

We should first note that species fixism is not an ancient Christian doctrine. Very few authors discuss prefixist transmutationism. Of these authors, many report on the fact that Thomas Aquinas, Augustine, and other church authorities such as St. Basil and Albert the Great (Aquinas's teacher) had categorically denied that God created all species during the first six days. Instead, God had conferred productive powers on various natural elements such as the earth and the waters (Zirkle 1951, 1959; Mivart 1871; Poulton 1908; Raven 1953). This power was thought to have produced various life forms at various times. The life forms can afterward reproduce their kind. However, the productive power is also active in the spontaneous generation of kinds in the present day, including such complex forms as eels, frogs, and mice. These theological opinions corresponded with contemporary secular authorities. The adoption of species fixism by religious authorities, and its reinterpretation as a "literal" reading of scripture, occurred only after its adoption by secular naturalists in the mid-eighteenth century.

Spontaneous generation is one version of antifixism. Another involves the transmutation of an individual during its lifetime. An example of this concept is butterfly metamorphosis, in which an individual organism is seen to transform from a worm-kind into an insect-kind. Less dramatic transformations were known, as when plants modify their form when they encounter a new climate. If these acquired modifications are conveyed to offspring (as everyone assumed they would be), then indefinite amounts of modification were possible.

Other kinds of transmutation occur across generations. Hybridization is one example. The giraffe was thought to have arisen from a camel–leopard pairing, and the hybrid origins of other species were unquestioned (Zirkle 1951). Other cross-generational transmutations were thought to occur during reproduction ("generation"). It has been known since antiquity that cultivated fruit do not reproduce their kind by seed. In the wild they do reproduce solely by seed, and so it was quite reasonably assumed that they do not reproduce true to type in the wild any more than they do in cultivation.

The belief in sudden, large mutations was widespread. In the thirteenth century, Albert the Great carefully described five ways that plants could change their species (Raven 1953: 70). "Peter Crescentius, the great fourteenth century agriculturist, devoted three chapters to sudden species changes . . . and for the next 200 years the sudden mutation of species was recorded in practically every work on natural history" (Zirkle 1951: 48). Francis Bacon believed "not only that one species might pass into another, but that it was a matter of chance what the transmutation would be" (Poulton 1908: 54). An especially common belief was that climate could permanently modify the species of plants; rye changed into cornflower, wheat and flax into other species (Raven 1953). John Ray claimed in 1687 that "Wheat . . . degenerates into tares, rape into radish . . . maize into wheat" (quoted in Crombie 1994 v. 2: 1270).

Ray's use of the term *degenerate* requires explanation. Degeneration in that age and context merely means transmutation, a process in which generation produces a form other than the parental form. Only later did degeneration come to imply deterioration. Zirkle believes that the modern failure to recognize pre-Linnean transmutationism is due to this semantic quirk. We mistakenly interpret early discussions of degeneration to imply degradation when no such implication was intended (Zirkle 1951: 48).

Even more dramatic transmutations were commonly accepted. To the modern ear they strain the boundary between myth and honest empirical belief. The story of the phoenix was often treated skeptically, but it was no less extreme than the barnacle goose. The *Oxford English Dictionary* still contains the renaissance term *anatiferous:* "producing ducks or geese, that is producing barnacles, formerly supposed to grow on trees and dropping off into the water below, to turn into tree-geese" (Hacking 1983: 70). Philosopher Ian Hacking uses the term *anatiferous* to illustrate incommensurability: What in the world could those people have been thinking of? But this was an honest factual belief. Raven quotes the sixteenth-century author Scaliger, who reports "as a thing he himself has seen" the stories "falsely told of the Phoenix but veraciously of the Bernacle [sic] Goose" (Raven 1953: 204).

Seen in the context of prefixist theories of spontaneous origins and transmutations, species fixism was a progressive scientific development. Beliefs in spontaneous generation persisted into the nineteenth century, but they were restricted to smaller and smaller organisms as time passed (Roe 1981). Fixism was established for nonmicroscopic plants and animals around 1750, primarily on the basis of plant-breeding experiments. Plant variation had been an especially common area of transmutationist beliefs. The careful and controlled breeding programs of Linnaeus and others established fixism among most naturalists. The importance of fixism as a scientific innovation was seldom acknowledged during the twentieth century. One exception is Charles Raven. In a discussion of the early years of the Royal Society (founded in 1660), Raven listed among its achievements the investigations that resulted in the law of gravity, the refutation of witchcraft, and the establishment of species fixism over spontaneous generation and transmutation (Raven 1953: 103).

Species fixism was important to the origin of evolutionary biology because it set the stage for the construction of the Natural System. If this is not obvious, consider the analogy with alchemy. A central aim of alchemy was the transmutation of base metals into gold. The alchemical tradition was not in its time considered magical; the "perfection" of base metals in the laboratory was believed to be an experimental recreation of what actually happened beneath the surface of the earth (Multauf 1966). Advances in chemical theory refuted the notion of transmutation of metals during the eighteenth and early nineteenth centuries. All substances were classified into elements and compounds, and the classifications have persisted through the modern day. The difference between elements and compounds is that elements cannot be transmuted, but they can enter into compounds in chemical reactions. Compounds are composed of elements, and they can be "transmuted" by decomposition in chemical reactions. Gold and other metals were established as elements. In this way, the alchemists' ambition to transmute base metals into gold was proven impossible. The chemical fixism of elements refuted the transmutationism of the alchemists. No one doubts that the periodic table and the identification of elements were major scientific achievements. Elements are fixed; they cannot be transmuted. It is a fact of nature.

But wait! The atomic theory of chemistry implies that elements are made of kinds of atoms, and that atoms are made of configurations of subatomic particles, and atomic reactions can rearrange the subatomic particles of any atom of matter. *So elements can be transmuted after all!* Does this mean that the nineteenth-century chemists were wrong, and the alchemists were right? Yes and no, but mostly no. It only means that the discovery of an important invariance in nature does not end the progress of science. That invariance may itself be discovered to hold only under certain conditions. Under the magnitudes of energy available in the nineteenth-century chemistry laboratory, elements could not be transmuted. The energy required to manipulate subatomic particles was far beyond the powers available to nineteenth-century chemists. The invariance is *real*, even though its true extent was unknowable at the time it was discovered. No alchemist deserves to gloat over the atomic transmutability of gold.

The same is true about eighteenth-century species fixism. Species fixism is *right*, and kind-mutationism is *wrong* about the changes undergone by organisms in the conditions known at the time, just as the chemists were right and the alchemists wrong about the transmutability of gold. To phrase the point in another way, both species and chemical elements are *natural kinds* with respect to the processes of change that were known in the eighteenth century: chemical reactions under moderate heat for chemistry, and breeding and environmental modification for species. However, they are not natural kinds with respect to processes of change that occur outside that range: atomic reactions for chemistry and geological time scales for species. Species fixism can only be shown to be wrong by considering time scales that are as far beyond the scope of eighteenth-century biology as the energies of a cyclotron are beyond the scope of the Bunsen burners of early chemical theory. No pre-Linnean transmutationist deserves to gloat over Darwinian evolution.

The discoveries of these two invariances were far from simple. In chemistry, the determination of which substances were elements and which were compounds was a tremendous task. In biology, the discovery of the sameness of insect metamorphosis but the nonsameness of maggots in rotting meat was as difficult. Linnaeus's belief in species fixity was supported by an intense program of exchange among horticultural gardens that demonstrated the reversibility of climatic change, together with the constancy of other characters that could be taken as diagnostic of the (fixed) species. In both chemistry and biology, the newly discovered invariance was accepted as universal for only about a century. Then it was broken. The invariance of chemical elements gave way to atomic physics; the invariance of species fixism gave way to Darwin.

Evolutionary theory does much more than simply deny fixism, of course – it explains things. Unlike prefixist transmutationism, the evolutionism of Darwin and all later thinkers presupposes a certain pattern of relationships among organisms, as well as very strong constraints on possible transmutations. Radical hybrids are ruled out, as are extreme changes of form (e.g., barnacles to geese, and worms to insects). The pattern of genealogical relationships that results from evolution is congruent with the pattern of systematic relationships among species. Without the recognition of systematic patterns among otherwise-unchanging species, evolutionism would have little to explain. This systematic pattern was constructed between the time of Linnaeus and Darwin. It was called the Natural System. Species fixism was necessary to the growth of the Natural System. Imagine trying to construct a coherent taxonomic system in which wheat could give rise to rye, worms to insects, mud to frogs, and barnacles to geese. The Natural System could not have been built without species fixism, and Darwinian evolution could not have been built without the Natural System. Evolution theory could no more have been discovered by a prefixist transmutationist than the Bohr atom could have been discovered by an alchemist.[8]

2.3 LINNAEUS AND HIS CONTEMPORARIES

Carl Linnaeus is important to our narrative for two reasons. He was prominent among the generation of naturalists who gathered the evidence for species fixism, and he produced the first widely accepted framework for the systematic classification of life.

---

[8] Ernst Mayr, founder of the Essentialism Story, acknowledges the widespread belief in transmutation before Linnaeus in a few references in his epic *Growth of Biological Thought*. He terms it *heterogony*, a term which I can find with this meaning nowhere else. He even acknowledges that Linnaeus's establishment of species fixism was progressive in somewhat the way discussed here (Mayr 1982: 259). This acknowledgment seems to have been lost in Mayr's own frequent advocacy of the Essentialism Story. Virtually every philosopher and most historians of biology with whom I have discussed pre-Linnaean transmutationism has been surprised (to the point of incredulity) that species fixism based on essentialism was not the dominant view of species prior to Linnaeus.

Linnaeus's fixism, like that of his contemporaries, was based on evidence that had been painstakingly gathered from a vast network of horticultural gardens across Europe. The old transmutationist beliefs in the influences of climate on plant forms had been tested by returning the modified forms to their original locations. The plants then reverted to their original forms. Experiments had been done in the production of hybrids ("bastards"), and the limitations on viability and fertility had made it seem exceedingly unlikely that this was a cause of new species. During this period, Buffon had gone against the emerging fixist consensus and argued in favor of "degeneration" (transmutation).

> [T]he best-informed naturalists found that Buffon had not made an adequate case for the concept of degeneration. The three principal external causes cited by Buffon for degeneration did not in observable cases change the form, proportion, or inner structure of an animal to a degree that would support the hypothesis.... Instinct and distribution kept the species pure; occasional crosses between species were sterile or soon reverted to the parental form. (Larson 1994: 84)

Linnaeus does speak of the essence ("essentia") of all plants, that is, of members of the plant kingdom. The essence of planthood is "fructification," the mechanism of generation in plants. Linnaeus worked out a complex physiological theory in order to account for the abilities of the sexual and fruiting parts of a plant to produce the embryo in the seed (Stevens and Cullens 1990; Müller-Wille 1995). Like most other generation theories of the time, Linnaeus's theory assumes the fixity of species. Linnaeus also speaks of the "essential characters" of genera and species. However, he does not assert that fixism must be true *because* of the essential characters, as the Essentialism Story would have it. Species fixity is a fact of nature, not of metaphysics. We will understand the cause of species fixity in plants only when we better understand fructification, the process of plant reproduction that produces offspring to resemble their parents. Fixism was treated by Linnaeus and others not as a metaphysical necessity but as an empirical fact of nature. Buffon's hypothesis of degeneration was subjected to empirical criticism, not metaphysical refutation.

As a systematist, Linnaeus laid out the hierarchical system of classificatory ranks and the binomial nomenclature by which a species is named. Species names were memorable but arbitrary, and they were not intended to allow identification of the species. Linnaeus himself called them "trivial names." His classification system for plants was based on a description of the configuration of sexual characters. The system was acknowledged to be "artificial" in that organisms were categorized on the basis of only one set of characters. Because Linnaeus acknowledged his system's artificiality, the lure of a truly *natural* classificatory scheme was great. The systematists who followed were all in pursuit of a Natural System. The quest for the Natural System may have begun with the innovation of species fixism, but it resulted a century later in a rejection of fixism – or at least a move beyond the particular sort of fixism endorsed by Linnaeus.

2.4   FRENCH SYSTEMS: JUSSIEU AND CUVIER

French systematists were in the forefront of the early quest for the Natural System. This work formed the basis of what I call taxonomic nominalism. Beginning in the 1770s, the French botanist Antoine-Laurent de Jussieu produced classifications that he considered natural because they were empirically based on a wide range of characters, instead of a restricted set such as Linnaeus's sexual system. By "natural," Jussieu did not mean that his arrangement reflected objective reality. In his view, no arrangement could do that. He accepted the principles of plenitude and continuity, principles quite at odds with essentialism (Lovejoy 1936). According to these principles, all possible kinds of organisms exist. Jussieu did construct hierarchical arrangements, like Linnaeus, but he did not believe that the hierarchy of his arrangements reflected a real hierarchical arrangement of organisms in nature. The lines of separation between adjacent groups are arbitrary. Adjacent groups flowed continuously into one another, so that no strict dividing line was dictated by nature to the observer (Stevens 1994: 75). Jussieu's nominalism about taxonomic groups was not based on his skepticism about the knowability of objective reality. Instead it was based on the belief that *there was no such reality*. With no objective structure to mirror, taxonomic decisions were based on pragmatics alone. For example, groups were determined on the basis of size, with each category containing between 2 and 100 members. The hierarchical arrangement and the requirement for at least two members in a group were justified by the fact that it allowed "generalization" in defining group characters. Characters that are carefully described for higher taxa need not be repeated for the lower taxa they included. Jussieu tried to give strict definitions of groups in terms of characters, but he regularly failed – and he showed no particular distress at the failure. Because nature itself was not hierarchically structured, the taxonomic hierarchy was merely pragmatic. Jussieu's methods are clearly inconsistent with the Essentialism Story as applied to taxa above the species level.[9]

Jussieu's views on classification and continuity were influential on his colleagues, significantly including Lamarck and Cuvier. Lamarck differed from Jussieu on the geometry of the resemblances among species. Lamarck's transmutationism assumed a continuous linear classification of all animals from lowest to highest in the tradition of the *scala natura*. By the turn of the nineteenth century, Jussieu's belief in overall continuity and Lamarck's belief in linear continuity were under challenge. Cuvier and his students were documenting the gappiness of nature. Cuvier introduced a four-part discontinuity at the most basic level of animal classification. Animals were portioned into four *embranchements*: Vertebrata, Molluska, Articulata, and Radiata. This arrangement was widely accepted. It is often said to have been based on the distinct structural plans of the four phyla, with Cuvier's embranchements interpreted as a version of the structural types discussed by von Baer and later morphologists (Russell 1916; Coleman 1964). However, this is a misunderstanding of Cuvier, and it is an important one (Winsor 1976; Ospovat 1981; Appel 1987). Cuvier understood the embranchements as distinct modes of functional organization, not distinct structural patterns. The similarities that were shared by members of an embranchement merely reflected their common functional needs. "For Cuvier, animals shared similar basic plans only because they carried out a similar combination of interrelated functions. . . . The embranchements were absolutely distinct from one another because the functional requirements of the animals in each embranchment were radically different" (Appel 1987: 45).

The concept of four *structural* types is very different from Cuvier's functionally defined embranchements, even though it placed organisms into the same categories. Karl Ernst von Baer and others proposed structural types in the 1820s. The structuralist, morphological approach is the topic of Chapter 3. I note it here only to insist on its contrast with Cuvier's functionally specified definition of the embranchements.

Cuvier considered the embranchements to be separated by gaps so great that no meaningful comparisons could be made between species of different embranchements. So he was a realist about the embranchements: they were distinct kinds of functional organization, not mere pragmatic groupings of organisms. Like most of his contemporaries, he was a species fixist, and so a realist about the species category. How about the intermediate taxonomic

---

[9] The discussion of Jussieu follows Stevens (1994; 1997).

ranks of genus, order, and class? Can we find evidence of essentialism or typology here?

In a sense, we can. Cuvier made use of what he called types in defining the intermediate taxonomic categories. Several concepts of *type* were current in this historical period, and some are reasonably interpreted as essentialist, but Cuvier's use of types was quite contrary to a Platonic use. He uses types to resolve the difficulty that we already noticed in Jussieu: Taxonomic groups were not generally definable by characters that were constant within the group. For an essentialist, this is unthinkable. For Cuvier and Jussieu, it was a mere pragmatic problem; the commitment to continuity made fuzzy group boundaries perfectly acceptable.

Cuvier's use of types has been called "classification types" (Farber 1976) and "the method of exemplars" (Winsor 2003). It is no comfort to the Essentialism Story. Exemplary types (as I call them) are used *specifically because of* the known impossibility of essentialist definitions of groups. The method works by choosing one member of a group (a genus within a family, or a species within a genus) as the type and then describing it very carefully. Nontype members are described only by their variations from the type. Both the method of exemplars and the choice of individual exemplary types are justified by pragmatic considerations only. The method eliminates unnecessary repetition and wasted ink. In 1828, Cuvier explained his pragmatic choice of the genus of the perch as typical for its family was merely because "it is a fish that is easy to procure" (quoted by Eigen 1997: 203).

Cuvier's arrangement was hierarchical, but his concept of the pattern of similarities was not. He conceived of the objective relationships among species as similar to a network or a fabric. "Cuvier conceived of an image of nature arising from the typical fishes situated as points in an otherwise continuous fabric. The fabric was patterned by regions of conformity centered on [typical genera] from which trailed lesser degrees of conformity until the next type was encountered" (Eigen 1997: 207). The choices of a taxonomic hierarchy and exemplary types had nothing to do with metaphysics, essentialist or otherwise. The decision was based on efficiency in data management. Cuvier's system encodes a very great deal of information about species in an efficient fashion. Natural categories differ from artificial ones not in revealing a hidden structure, but in summarizing large amounts of data in an efficient way (Coleman 1964: 187). However, the data are always *about individual species*, not about real relations between species, and certainly not about real taxonomic entities such as genera and families. Cuvier did not distinguish between deep and important similarities (affinities, later called homologies) and superficial similarities (later called analogies). His nominalism about the

43

intermediate taxonomic ranks (class, order, and genus) is conceptually tied to his refusal to distinguish affinities from analogies. Similarities are similarities, threads in the continuous fabric of life.

Cuvier's deepest commitment was to functionalism: The deepest facts about the organic world are functional facts. The nature of organisms was to be understood as a consequence of the *conditions of existence*: In order to exist at all, an organism must possess the kind of internal organization that allows it to fulfill its physiological needs, in the environment it lives in. Cuvier claimed that the strict demands of functional integration among the various parts of an organism made it possible for the paleontologist to infer the nature of an entire organism from the discovery of only a part of one bone. He even proposed a functionalist explanation for the fixity of species. The functional integration of parts was so finely balanced in a species that a change in any part would make the species nonviable. Transitions between species simply could not exist.

> Every organized being forms an entire system of its own, all the parts of which mutually correspond, and concur to produce a certain definite purpose by recip-rocal reaction, or by combining to the same end. Hence none of these separate parts can change their forms, without a corresponding change in the other parts of the same animal; and consequently each of these parts, taken separately, indicates all the other parts to which it has belonged. (Cuvier 1813: 90, quoted in Whewell 1863 v. 2: 493)

Cuvier's functionalist account of species fixism is widely recognized (e.g., Gould 2002: 295). His taxonomic nominalism is not widely recognized. Nevertheless, the two concepts are closely related. Cuvier believes that all species that *can* exist *do* exist. This meshes nicely with his nominalism about taxonomic groups. Any combination of characters that exists, exists because it can, not because it fits into the structure of the Natural System. Genera, classes, and families have no independent reality; they are human concoctions. Each species is "an entire system of its own." His fabric-like image of species similarities leaves each species as a self-subsistent individual.

Cuvier's species fixism and his taxonomic nominalism are equally contrary to an evolutionary concept of life. Species fixism could not give way to a modern concept of evolution until taxonomic nominalism was weakened. Before species can be considered to have common ancestors, they must first be considered to have real relationships. Within systematics, that trend began in England.

2.5 BRITISH SYSTEMS AND THE GROWTH OF TAXONOMIC REALISM

By the 1820s, four distinct "natural systems" of classification were being discussed in Britain. Two were already discussed: Cuvier's embranchements and hierarchical system, and Lamarck's system of linear progressionism. Two more systems originated on British soil. One was the dichotomous system devised by Jeremy Bentham in his reformist educational tract *Chrestomathia*, published in 1817 (Bentham 1969). The system was based entirely on Bentham's empiricist epistemology, and it made no commitment to the objective structure of the domain being analyzed. The other was William Sharp MacLeay's circular system, often called the *quinarian system*. Its full-blown version is baroque and quaint, and it has received a great deal of somewhat incredulous attention from historians. Nevertheless, it made an important and influential step toward taxonomic realism.

MacLeay's original system was introduced 1819. Its importance for taxonomic realism comes from MacLeay's claim that he had detected two distinct kinds of similarity among organisms, which he termed *affinity* and *analogy*. The distinction derived from the peculiar geometry of the system. First, natural affinities (the similarities that mark the closest natural relations) connect species in a linear fashion. This point is consistent with Lamarck. Unlike Lamarck, MacLeay claimed that the closest affinities bound together relatively small groups; linear relations did not stretch across the animal kingdom. Second, each affinity group ran parallel to other affinity groups at the same rank. The parallelism was constituted by the second set of similarities (analogies) that connected members of affinity groups at corresponding points along the linear sequences. So far, a ladder-like geometry is implied. But lastly, MacLeay believed that the individual chains of affinity closed at the ends to form circles. Parallelism and closure together imply that the same number of members must be present in each circle. MacLeay suggested that the number would be the same at all taxonomic levels, and proposed five as the universal taxonomic number. In fact, he had little commitment to fives and did not stress the universality. The important aspect to MacLeay was his "discovery" that *two distinct sets* of similarity relations (affinity and analogy) revealed the organization of life (Winsor 1976: 82 ff.).

MacLeay left London in 1826. His theory was picked up and elaborated by William Swainson and subsequently by the anonymous author of the *Vestiges of the Natural History of Creation* in 1844 – thence derived its notoriety and its quinarian commitment to circles of five members. Serious naturalists preferred MacLeay's original version. T. H. Huxley began his scientific career while traveling as an assistant surgeon on the scientific ship *Rattlesnake* in the late 1840s. He met MacLeay in Sydney, and he was very impressed with his ideas. He complained that, up until then, all he had known about MacLeay's thought had come from "Swainson's perversions" (Winsor 1976: 87).

The importance of MacLeay's innovation is his taxonomic realism about the affinity groups. In contrast to the nominalism of Jussieu and Cuvier, MacLeay thought that he had found the objective pattern of organic relationships, and it lay in a distinction between "deep" or "real" similarities and similarities that were more superficial. "It is quite inconceivable that the utmost human ingenuity could make these two kinds of relation to tally with each other, had they not been so designed at the creation. A relation of analogy consists in a correspondence between certain parts of the organization of two animals that differ in their general structure" (MacLeay quoted in Winsor 1976: 85). It is very hard for modern thinkers to take the circular system seriously, of course. MacLeay's commitment to a regular geometric pattern and his allusions to divine creation have both been disparaged. I certainly agree that MacLeay was extravagantly mistaken. However, complaints about his allusions to creation are misplaced (as such complaints often are when made by modern commentators on pre-Darwinian authors). His reference in the aforementioned quotation to the design of creation is merely his way of insisting that the circles are *real*, not a human convention. Unlike Cuvier, MacLeay believed that he could discern real, objective affinities among groups of organisms. His realism about the affinity groups had strong influences on naturalists such as Huxley, who dismissed his circles and universal numbers but continued to delve into the realistic basis of taxonomic relationships.

The popularity of the four natural systems resulted in an intriguing debate among naturalists in England during the 1830s. It concerned, of all things, the naming of species (McOuat 1996). Establishment naturalists were Linnaeans, and they believed like Linnaeus that species names should be arbitrary markers, conferred by qualified experts. Dissidents argued that species names should reflect the position of the species within the Natural System (even though the nature of that system had not been settled). According to McOuat's fascinating report, the debate "rumbled on for years" in scientific societies. Its resolution came about through the efforts of Hugh Edwin Strickland, a young Oxford-educated naturalist. Strickland argued for the Linnaean convention on the basis of John Locke's theory of language. Names are conventional signs for things, and they need not be descriptions. Systematic theories are subject to change. To link species names to the vagaries of systematic theory would invite linguistic chaos; all museum specimens would have to be relabeled every time the systematic theory changed.

46

Not content with argument, Strickland wanted to institutionalize his Lockean view of species names. He published a provisional set of Rules for Zoological Nomenclature, and he convinced the British Association for the Advancement of Science (BAAS) to form a committee to consider the rules. The committee included Strickland, Charles Darwin, Richard Owen, and other prominent naturalists, and questionnaires were widely distributed. All of the committee members were opposed to the radical systems, and Lockean–Linnaean naming conventions won the day. Strickland managed to get an agreed-upon set of rules published in the 1842 BAAS Report.

The most striking feature of the rules is what they did not say. They set down a process for naming species, and they explained how names (unlike descriptions) arbitrarily signify things; species names designate species. What about the species concept itself? The rules were silent. Nothing was said about fixity, or even the objective reality of species. This was not Strickland's decision; it was the desire of his many commentators. Strickland had modestly proposed to codify species (in contrast to genera) as "tangible objects." Even this wouldn't fly, with one commentator asserting that species and genera alike were mere abstractions. "The recorded changes made to the drafts at the meetings at the Zoological Society . . . and as a result of the voluminous correspondence with British and international naturalists, all point in one direction: *against an ontological and definitional commitment to species*" (McOuat 1996: 512; emphasis in the original).

So the species category was not defined in a fixist manner, or in a creationist or essentialist or typological manner. It was not defined at all. Species were nameable somethings.[10] Not only do we find no evidence of essentialism in a place that it might have appeared: We don't even find evidence of species fixism!

Strickland's work on nomenclature was conservative. He disapproved of the a priorism and symmetry of the radical systems. However, he considered MacLeay's distinction between affinity and analogy to be crucially important. He elaborated on it, in a way that contributed even more to the reification of the Natural System. Strickland's views on affinity and analogy were first articulated in an 1840 paper, written in response to a publication that had claimed that the distinction between affinity and analogy was merely a matter of degree. Strickland disagreed. Affinity and analogy were different in kind.[11]

---

[10] John Beatty has attributed this noncommittal approach to species to Charles Darwin, as a technique to avoid philosophical tangles in the *Origin* (Beatty 1985). McOuat has shown that the technique was already in place within Darwin's circle of naturalists.

[11] This moves Strickland beyond MacLeay, who continued to believe that both affinity and analogy were objectively real relationships in nature. Huxley also disagreed with MacLeay on this issue.

Affinity is a similarity that marks a real, natural relationship between species or groups. Analogy is a secondary and accidental similarity, and it is no indication of relatedness. Natural relationships are those that reflect the position of a group within the Natural System.

> [I]f this [adaptation] were the *sole* mark of design, if each species constituted a being *per se*, adapted to its peculiar condition of existence, but not allied in physiological structure to its fellow species, there would then be no *natural system* . . . there would be none of those essential peculiarities of structure which we find to pervade vast groups of beings whose external forms are often widely dissimilar. The existence then of a comparatively few grand types of structure . . . may be taken as a proof that species were created not absolutely, but relatively. (Strickland 1840: 220; emphasis in the original)

These statements are an articulate expression of the ontological contrast between taxonomic nominalism and realism. Taxonomic nominalists, including pre-Darwinian adaptationists such as Cuvier and the *Bridgewater Treatise* authors,[12] considered every character of an organism to be adaptive. A character's significance was its function *for that species*. As Cuvier stated here, each species "forms an entire system of its own." Each species exists only per se, as Strickland puts it. Realism about the Natural System requires one to recognize certain characters as marks of a genuine relatedness within the structure of the Natural System. This is what it means for species to be created "not absolutely, but relatively."[13] Strickland goes on to explain how his concept of affinity is tied to the real existence of a natural (as opposed to an artificial) system.

> [W]e may proceed to define *affinity as the relation which subsists between two or more members of a natural group*, or in other words, *an agreement in essential characters*. . . . Hence we see why the idea of a *natural system* is necessary to the definition of *affinity*, for in an *artificial system* the characters of the groups are not *essential*, but *arbitrary*, and the relation between the members of such a group would be, not *affinity*, but mere *resemblance* or *analogy*. (Strickland 1840: 221; emphasis in the original)

As an example of an artificial group, Strickland offers a definition of *Pisces* in which the group is defined in terms of adaptation for swimming. This unnatural group would include whales and porpoises. The similarity in outward

He suggested to MacLeay that affinities were based on developmental commonalities, whereas analogies were based on adaptive convergence (Winsor 1976: 91).

[12] The *Bridgewater Treatises* are discussed in Chapter 3.

[13] Those who are still squeamish about the language of creation should get over it. Strickland's term *created* should be read as "come into existence"; he is writing science, not natural theology.

form of fish and whales is mere analogy, dictated by functional adaptation: "Analogy, in short, is nothing more than *an agreement in non-essential characters*, or a resemblance that does not constitute affinity" (Strickland 1840: 222; emphasis in the original).

Strickland's realism about groups is strikingly illustrated in a passage about the interpretation of anomalous species, those that seemed to have characters of more than one group. Cases such as the platypus and hagfish were no particular problem for advocates of continuity; those species sat squarely on the blurred line between groups (birds and mammals for the platypus, and fish and mollusks for the hagfish). They also fit well enough into Swainson's elaboration of quinarian theory in which "osculant" species or groups were positioned at points of contact between two circles. But Strickland took the discrete integrity of classificatory groups very seriously. The Natural System had a determinate structure of real relationships, and analogies were no part of them. Affinities were real links, and analogies were fortuitous similarities based on similarities in adaptation.

> Thus if we suppose *all birds* to be equally distinct in essential structure from *all mammals, all Vertebrata* from *all Molluska*, it is plain that the *approximation* between *Ornithorynchus* [duck-billed platypus] and birds, and between *Myxine* [hagfish] and *Molluska*, resolves itself into mere analogy. But if birds have a *tendency* to unite with *mammals* by means of *Ornithorynchus*, and *Vertebrata* with *Molluska* by means of *Myxine*, then this *approximation* is not to be considered as a distinct principle, but only as an undetermined analogy or affinity. (Strickland 1840: 226; emphasis in the original)

This passage expresses two important ideas. The first is at odds with Cuvier and other nominalists: Strickland's rigorous realism about class-level groups. Strickland says that "all birds" are equally distinct from "all mammals." Taxonomic nominalists, along with anyone who accepts continuity among species, can see that some birds are more mammal-like than others, and some mammals are more birdlike than others. How can all birds be equally distinct from all mammals? They can be equally distinct *only* if class-level taxonomic groups such as "bird" and "mammal" are real entities in which group membership is determined by essential characters, not overall similarity or conventional decision.[14] The second point is that Strickland's realism about the Natural System is tied to fallibilism about its discovery. He is even willing

---

[14] Cuvier did indeed claim that "all mollusks" are equally distinct from "all vertebrates," because they are distinct embranchements. However, Cuvier was a conventionalist about intermediate taxa. In addition, his commitment to real embranchements was based on functional categories, so no ontological commitment to groups was needed.

to imagine the discovery that vertebrates and mollusks – members of distinct embranchements – actually belong to the same group.

Finally, we must recognize that Strickland offers the first use of essential-ist vocabulary that we have seen. Essential characters are those that reveal affinities, and thereby place a group into its correct position within the Natural System. I submit that this is a *programmatic definition* in Boyd's terms (see Section 1.5.3 in Chapter 1). It shows the organization of the Natural System, but without providing any "ultimate" explanation of its structure. Individual species are members of groups *because* they share the groups' essential characters. Unlike the stereotyped Platonic essentialist, Strickland is a fallibilist and an empiricist about what these characters actually are. It has been claimed that essentialism (or typology or idealism) makes it impossible to conceive of real-world connections or intermediaries between distinct types (Bowler 1996: 43). Strickland shows the mistake in that view.[15] I will argue that this kind of essentialism about taxonomic characters and realism about the Natural System, far from being a barrier to evolutionary thinking, was a crucial contributor to it. We have finally found something that could be called essentialist. It turns out to be based not on a prioristic Platonic nonsense but on good British empiricism. It's scientifically progressive to boot.

2.6 REVIEW OF SPECIES FIXISM, ESSENTIALISM, AND REAL GROUPS

Let us review the cases discussed so far with respect to the related issues of species fixism, essentialism, and realism with respect to the Natural System.

Species fixism was an innovation of the mid-eighteenth century. It rejected a chaotic collection of transmutationist beliefs and thereby made it possible to construct a Natural System. Linnaeus's belief in fixism was based primarily on the empirical work of botanists, not on metaphysical grounds. His use of the language of essentialism occurred in contexts that are impossible to map onto species fixism. Like many other naturalists of his age, he was fascinated with the problems of generation, and he spoke of the "essence" of plants as their reproductive powers. He also spoke of essences of species and genera, but he apparently took them to be merely those constant characters that would allow us to taxonomically identify the plants. Linnaeus's essentialism *followed* his species fixism, rather than being the grounds of it.

---

[15] Others such as George Waterhouse shared Strickland's taxonomic essentialism (Waterhouse 1843). Darwin corresponded with Waterhouse, recognizing the importance of the view that anomalous species such as the platypus will eventually be located within natural taxa (Burkhardt and Smith 1986: 415).

Jussieu believed in continuity between groups. His classifications failed to pick out necessary and sufficient conditions for group membership, and they were heavily influenced by pragmatic criteria. Cuvier shared Jussieu's belief in continuity and produced similarly "polythetic" groups. Both Jussieu and Cuvier were species fixists but taxonomical nominalists. They considered species to exist per se, as distinct entities, not as members of higher taxa. Cuvier used the method of exemplary types, which is a method clearly inconsistent with essentialism (Winsor 2003). For these authors, essentialism was clearly false of genera and higher taxa. We have no evidence that Jussieu's species fixism was based on essentialism or anything other than Linnaeus's empirical evidence. Indeed, in Cuvier's case, his functional explanation of species fixism leaves essentialism redundant.

MacLeay was certainly a realist about the existence of groups, and also about the group-crossing relations of analogy. We have seen no specific claims of essential characters, and so I see no direct application of essentialism to this case. There is a tendency to identify MacLeay as an idealist (based on his so-called ideal circles), to infer that idealists are essentialists (perhaps because Plato was both idealist and essentialist?), and to conclude that MacLeay must have been an essentialist. I reject the inference, but the point is moot. I am not trying to duck the essentialists. Essentialists are about to appear.

Strickland was a genuine essentialist, though quite different from the Platonic stereotype. Essential characters are those that *make that species what it is* with respect to its taxonomic position: its phylum, order, class, family, and genus. Strickland gives no hint of conventionalism about these groups, as did the French systematists and the British dichotomists. His essentialism about groups is tied to an empiricism about our knowledge of groups, however. What does Strickland's essentialism about groups imply about the fixity of species? I see no implication at all. He may have been a fixist, but we have no indication that he was an a priori fixist, that species change was inconceivable to him. We have no reason to doubt that Strickland was a responsible empiricist about species fixism, just as he was about the objective reality of the Natural System. One final point about Strickland's essentialism: It was apparently tied to classification alone. He gave no indication that the discovery of taxonomic essences would enable new causal explanations. In this way, his classificational essences were merely extensions of Linnaeus's species essences. They are the true basis of the Natural System, the organization of life. However, Strickland gives us no hints of what he would consider a deeper explanation of the Natural System. The search for deeper laws underlying the taxonomic regularities is seen in Chapters 3 and 4.

Within our range of systematists, we see no confirmation of the claim that species fixism was grounded in essentialism or typology. I suggest that the *most* antievolutionary systematists were the the taxonomic nominalists. Essentialism, when it appeared, was in support of taxonomic realism. For this reason, essentialism was associated not with antievolutionary thought but with the kind of thought that would enable Darwin to argue for descent from common ancestry. Practitioners of Synthesis Historiography have commented on Strickland's views. It is said, for example, that Strickland defined neither "essential" nor "natural system" in his distinction between affinity and analogy (Mayr 1982: 209). True enough. However, if this is intended as a criticism (as it appears to be), it is no more appropriate than criticizing Kepler for failing to explain what kept the planets in elliptical orbits, that is, criticizing him because he was not Newton. I believe that these steps toward *real groups* were in fact progressive.

More essentialists are to come, and they will play an important role in the development of evolutionary thought. However, the ideas will not come primarily from systematists. Both the traditional distinction between natural and artificial classifications and my distinction between realist and nominalist interpretations of taxonomy oversimplify the situation within systematics (Scharf 2003). Classifications were devised for too many different purposes to unequivocally support the kind of thought that would lead to evolution. The truly important concepts would arise not from systematics but from morphology and embryology. To these studies we now turn.

3

The Origins of Morphology, the Science of Form

Morphology is the study of organic form. It began as a branch of systematics, but it grew into one of the central biological disciplines of the nineteenth century (Nyhart 1995). In response to Darwin's *Origin*, morphology became one of the two distinct and enduring approaches to the study of evolution, the other being neo-Darwinism itself. The basic research program of morphology will follow our discussion throughout the book, even to the modern day.

In this chapter I have two goals. First, I sketch the origins and early nature of the research program of morphology. Second, I discuss and evaluate how practitioners of SH have dealt with pre-Darwinian morphology. One particular interpretation has been extremely influential. Peter Bowler in 1977 argued that pre-Darwinian British morphology should be seen as an innovative version of natural theology, in particular an "idealistic version of the Argument from Design" (Bowler 1977). This interpretation immediately caught on, and it has been accepted not only by authors with an SH orientation (Ruse 1979, Mayr 1982) but also by those with a structuralist slant (Ospovat 1981; Gould 2002). Nevertheless, I believe the interpretation to be importantly misleading. My (revisionist) understanding of the history of evolutionary theory requires that we recognize a deep division between the pre-Darwinian natural theologians and the morphologists. If we see morphologists as natural theologians, we will be unable to recognize the importance of their contributions to evolutionary theory. Moreover, we will be unable to recognize the historical depth of the research program that has led to evo–devo.

Let us first briefly consider the history of the Argument from Design (AD). Natural theology is the use of ordinary empirical evidence in support of religious beliefs. The centerpiece of natural theology is the AD. In the nineteenth century, the AD was based entirely on empirical evidence of the means–ends

53

adaptations that were exhibited by plants and animals. Premises about adaptation, together with the principle that adaptation can only be created by an intelligent mind, were taken to imply the conclusion that an intelligent mind created the organic world. However, earlier versions of the AD were not based on adaptation alone. In ancient Greek times, both organic adaptations and lawlike patterns of the movement of the heavens were used to argue for the creation of the world by an intelligent designer. Plato proposed a Paley-like version of the AD, arguing that an intelligent creator was the best answer for the purposive question about the world, "Why is it good?" Stoics inferred a creative intelligence not from goodness but from nonpurposive patterns. The regularity of celestial motions was likened to the motions of a fleet of ships moving across the horizon (Hurlbutt 1965: 108; Glacken 1967: 56). Just as we would infer a pilot of the fleet, we would infer an intelligent creator of the heavens, even if we were unable to discern the purpose behind the pattern. The equivocation between purpose and pattern continued throughout the middle ages. Aquinas's fifth proof of God's existence equivocates between "always acts the same" and "acts to achieve the best results" as evidence for an intelligent designer (Aquinas 1952: 13).

Ironically, the separation of purpose from pattern was stimulated by Isaac Newton's scientific achievements. After heavenly motions were explained, they lost their argumentative punch. Around the turn of the eighteenth century, prominent natural philosophers such as John Ray and Robert Boyle admitted that astronomy could no longer bear the theological weight of the AD. The weight fell completely to biology, and the purposive adaptations that were continually being discovered by naturalists.

It is useful to distinguish primary from secondary arguments in natural theology. A primary argument demonstrates God's existence. A secondary argument does not prove God's existence, but it demonstrates his attributes (Paley 1809: 57). Astronomy had formerly comprised a primary argument for God's existence, but after Newton it was demoted to secondary status. William Paley acknowledged that astronomy "is *not* the best medium through which to prove the agency of an intelligent Creator" (Paley 1809: 378; emphasis in the original). The heavens still proclaimed the glory of God, but they had ceased to prove God's existence. All that remained of the AD was the Argument from Purpose (Gillespie 1987; Amundson 1996).

The nineteenth century brought a new challenge to the biologically based AD: the geological facts of an ancient earth and organic extinctions. A new generation of natural theologians, liberal for the time, was forced to battle against biblical literalism to establish modern non-Biblical geology in university curricula (Cannon 1978). The core of natural theology was revised from the once-only Genesis creation of individual species, with adaptations of each species fitting it into an unchanging environment, to the successive creation of species through geological time, with adaptations designed for the changing environments they were created within. The last Earl of Bridgewater in 1829 commissioned a set of eight treatises intended to document the continued validity of the adaptationist AD.

We will return to natural theology after we examine the continental development of morphology over this same period. Morphology itself is a pattern-like study. It does not identify goals or purposes for the morphological patterns that are discovered. At first glance, it would seem no better suited than astronomy as a source of evidence for intelligent design.

3.2 FORM AS A TOPIC OF STUDY

Like adaptation, the recognition of organic form and comparisons between forms can be found in ancient Greek philosophy. Unlike adaptation, organic form did not constitute its own area of study. The notion that structural correspondences should be studied, and not just noticed, was invented around the turn of the nineteenth century by Goethe in Germany and Geoffroy in France. As superficial patterns were recognized, deeper patterns emerged. Some repeated patterns were recognized within a single body, others came from comparisons of different species, and some came from the embryological study of how form arose during development of an individual. *Form* in its morphological meaning designates structure rather than mere shape. For this reason I will refer to *structuralist* rather than *formalist* biological approaches. Groups of organisms were observed to be built on similar structural plans, even though these plans were subject to great variation. Variations were often called *transformations* or *modifications*. These terms were not intended to indicate temporal evolutionary change, but variation among corresponding body parts.

3.2.1 Goethe

The nature of the first morphological correspondences differed in Germany and France. Johann Wolfgang von Goethe emphasized the fact that whole bodies are made up of repeated elements that sequentially corresponded with one another. Richard Owen later termed these correspondences *serial homologies*. Goethe discusses two classic cases. First was his conjecture–discovery in the late 1780s that the various parts of plants were all modifications of one basic part, most easily thought of as a primordial leaf. The second was an extension of the obvious fact that the primary axis of the bodies of vertebrates was almost entirely composed of repeated elements, vertebrae. The exception was the skull. However, the skull itself could be seen as a number of modified vertebrae, with the brain as an expansion of the spinal cord. This is the vertebral theory of the skull, for which credit has been commonly (but probably falsely) been given to Goethe.[16] Goethe's basic insight, that bodily form could be understood as repetitions of elements that had an underlying unity, was one important theme in nineteenth-century morphology. Goethe had also discovered an important morphological correspondence between species. This was the intermaxillary bone in humans, a small bone in other vertebrates that had been claimed to be absent in humans. Goethe discovered it in fetuses and young children. His commitment to the common general form of all vertebrates, which he termed the *archetype*, was the basis of this search (Appel 1987: 160).

3.2.2   The Great Cuvier–Geoffroy Debate

Geoffroy's morphology was similar in spirit to Goethe's intermaxillary discovery. It relied not on comparisons of parts within individual bodies, but comparisons between bodies of different species, what Owen would later call *special homologies* (with "special" referring to species). The obvious correspondences had always been recognized, such as heads and limbs in vertebrates. As comparative anatomy proceeded, the level of correspondence became increasingly detailed with new and finer correspondences. At first this study was an uncontroversial aspect of comparative anatomy. Geoffroy and his colleague Cuvier cooperated in the early studies. As discussed in Chapter 2, Cuvier expected to find corresponding body parts in organisms that were functionally similar. However, Geoffroy began to discover correspondences that clearly did not reflect function. His structuralism amounted to the belief that these structural correspondences existed independently of functional needs. This claim of the autonomy of structure produced a conflict with Cuvier's thoroughgoing functionalism. Geoffroy stated the principle of the Unity of Type, in opposition to Cuvier's principle of the Conditions of Existence. The clash between these two principles was recognized

---

[16] Robert Richards has argued that the *Naturphilosoph* Lorenz Oken actually devised the theory in 1807. Goethe claimed priority in the 1820s, but he apparently did so as a result of misremembering his experiences of the 1780s, when he had formulated the plant theory (Richards 2002: 477 ff.).

throughout the scientific world, and it was prominently discussed in Darwin's *Origin.*

The conflict polarized as time passed. Beginning in 1800, Cuvier published a massive study of the four classes of vertebrates, stressing functional differences between the classes. Geoffroy in 1807 began searching for structural similarities that did not correspond to similarities in function. An example was his treatment of the furcula bone. Cuvier claimed that the bone existed only in birds, where it served an important function in flight. Geoffroy identified the furcula in fish. Thus, in Appel's words, "the furcula was not a bone specifically designed by the Creator to aid birds in flight, but rather an abstract element of organization which could serve multiple functions as it was placed in different circumstances" (Appel 1987: 87).

Appel's phrasing illustrates an important point about idealism. The modern reader might wonder what is "abstract" about Geoffroy's identification of correspondences between bones in fish and birds. Bones are bones, not abstractions. Nevertheless, Appel's description is exactly appropriate to the context of the early nineteenth century. Observations of functionally individuated entities (e.g., "wings" in insects and birds) were regarded as empirically simple and direct. Observations of structural identities (i.e., those that did not reflect function) were regarded as inferential and epistemologically more suspect. The epistemological alignment of function with direct observation and conservative empiricism worked to the benefit of Cuvier and the British natural theologians for many years. When Geoffroy and others spoke about identities that were irreducible to function, they had no choice but to admit that they were abstracting from experience, and therefore speaking about "ideal" entities. I will refer to this epistemological policy as the *empirical accessibility of function.*

The modern antipathy toward idealism may be partly defrayed by recognizing this policy. Entities were considered abstract or ideal just in case they had been inferred from observation, rather than directly observed. Such entities had no special ideal metaphysical status. Functional identifications were taken as observable, whereas structural identifications (when generalized across species) were taken as abstractions or idealizations. Like other methodological standards, this one changes with time. By about 1850 the tables had turned. Structural facts had come to be treated as empirically unproblematic, and functions were seen as speculative and in need of special justification. In Geoffroy's day, though, his claims of the identity of organs could only be seen as abstract, merely by the fact that they were not based on function.

Geoffroy proposed homologies between increasingly diverse body parts. For example, mammalian ear ossicles, the tiny inner ear bones, were identified with fish opercular bones, gill covers. The conflict escalated in 1820. Three anatomists, including Geoffroy, presented homological theories that compared articulates and vertebrates – finally violating Cuvier's embranchements. Geoffroy proposed the "seemingly preposterous" homology between the exoskeleton of arthropods and the endoskeleton of vertebrates, proposing that "every animal lives within or without its vertebral column" (quoted in Appel 1987: 111). The problem was that the nervous system of vertebrates is on their ventral side, and that of arthropods on their dorsal side. Geoffroy replied that ventral and dorsal were relative terms; insects merely moved with their neural side down and their haemal ("blood-side") side toward the sun. Appel, writing in 1987, reports that "Such comparisons seemed no less fanciful to his contemporaries than they appear to us today" (Appel 1987: 111). As recent evo–devo discoveries illustrate, criteria for preposterousness change with time, and have changed radically even since 1987 (see Chapter 1, Section 1.2). The battle between Cuvier and Geoffroy became personal and public; it began in February 1830 with a series of six papers presented before the Académie des Sciences over a proposal of homologies between elements of the vertebral torso and organs of mollusks. Geoffroy was inspired to assert that the body plans of mollusks and vertebrates would eventually be reduced to a single type. The embranchements again were violated.

The outcome of the great Cuvier–Geoffroy debate was inconclusive. Cuvier seemed to have the advantage, at least in terms of public reputation, but he died two years after the debate. Traditional histories have awarded the win to Cuvier, but Appel and others point out that Geoffroy's structuralism (if not his iconoclasm) was to dominate the following decades.

3.2.3   Von Baer and Development

Two sets of structural patterns were already mentioned: the repetition of structural units within an individual body, and the correspondence of body parts in organs of different species. Each is a static pattern, in the sense that comparisons are made at a moment in time. The third aspect of morphology is dynamic: the pattern of changes in bodily form that occur as an individual organism develops from a single cell to an adult. This is embryology. Once we identify these temporal patterns in different species, we can compare them. We immediately detect patterns in the embryological sequences that show relations among species. The first pattern to be recognized was the law of parallelism, or the Meckel–Serres law, so named by Russell to acknowledge its earliest proponents (Russell 1916: 94). This law was based on a linear conception of the Natural System (the scala natura), and a correspondingly linear embryological trajectory. The Meckel–Serres law states that the embryological development of higher organisms involves a succession of stages that represent adult forms of organisms that are lower on the scala natura. The law was soon refuted (though not completely abandoned), but it does illustrate the morphological goal to explain organic form by organizing the vast variety of forms under general laws.

Karl Ernst von Baer thoroughly refuted the Meckel–Serres law, and in doing so made embryology a science (Russell 1916: 118). His embryological publications began in 1828. He summarized the results of his study with four laws of embryological development:

1. That the general characters of the big group to which the embryo belongs appear in development earlier than the special characters.
2. The less general structural relations are formed after the more general, and so on until the most special appear.
3. The embryo of any given form, instead of passing through the state of other definite forms, on the contrary separates itself from them.
4. Fundamentally the embryo of a higher animal form never resembles the adult of any other animal form, but only its embryo. (von Baer 1828: 224, quoted in Russell 1916: 125–126)

Like many others in the early nineteenth century (including Cuvier and Lyell), von Baer considered evidence against the linearity of the Natural System to be evidence against evolution. And so it was, at least against Lamarckian versions of evolution. Von Baer described the embryo as progressing from an originally homogeneous, undifferentiated state to progressively more heterogeneous and differentiated states. Embryos of one species resemble embryos of other species, and they do so in a temporal pattern that reflects their relationship in the Natural System. Von Baer did not consider the Natural System as a linear scala natura, but a hierarchy of groups within groups. Embryos of distantly related species resemble each other only very early in their development. Embryos of a closely related species resemble each other until much later.

The chick is first a Vertebrate, then a land-vertebrate [i.e. not a fish], then a bird, then a land-bird, then a gallinaceous bird, and finally *Gallus domesticus*. (Russell 1916: 125)

Von Baer divided the animal kingdom into four types that nominally corresponded with Cuvier's embranchements, but his concept of the cause of the unity within a type was crucially different from Cuvier's. To von Baer, Unity of Type was an effect of shared patterns of embryological development, what he called schema of development.

> In reality, instead of a "Type" and a "Schema" I might have used a common term expressing both. . . . The schema of development is nothing but the becoming type, and the type is the result of the scheme of formation. For that reason the Type can only be wholly understood by learning the mode of development. (von Baer 1828, quoted in Lenoir 1982: 86)

The type is unified by its mode of development, not (as Cuvier would have it) by the functional needs of the adult organism. This is a very radical difference. Whereas Cuvier had explained adult similarities by similarities of (adult) function, von Baer explained them by their mode of embryological generation, completely ignoring adult function. This contrast would persist through the next two centuries. Von Baer's importance to our story is in his laws of divergent embryological development. However, we should recognize one other factor. Von Baer was a teleologist, but a teleologist of a kind quite distinct from the natural theologians. He was what Timothy Lenoir has labeled a *teleomechanist* (Lenoir 1982; Larson 1994). Teleomechanists were Kantians with respect to teleology, regarding teleology as a regulative or heuristic principle for biology. The paradigmatic case of teleology for Kant and the teleomechanists was embryological development. Early embryological stages exist in order to developmentally produce adult stages (which then reproduce new embryos). Early embryonic stages are necessary causal conditions for later stages, but not a mere series of efficient causes. Nevertheless, we can gain what I will call a quasi-causal understanding of adult form when we recognize how it arises out of embryogenic processes.[17]

It is important not to confuse the teleomechanists with the Paleyan natural theologians. Both were teleologists, but their differences are more important than their similarities. The paradigmatic case of teleology for Paleyans was the same as for Cuvier: the adaptation of an adult character to an environmental need. Von Baer and the teleomechanists were on the structuralist side of the function–structure dichotomy; the Paleyans and Cuvier were on the functionalist side. Like idealism versus realism (or idealism versus materialism, or

---

[17] This brand of teleology seems never to have played a role in the British debates; it was not appealed to by natural theologians, and it was not opposed even by such antiteleologists as Thomas H. Huxley, who himself translated von Baer's work into English.

whatever), the philosophical contrast of teleology versus nonteleology does not capture the nature of this dispute. Function versus structure does.

3.2.4 The Study of Form Summarized

Let us review morphology and its goal of explaining form. Form was studied not merely as individual static shapes, but as dynamic and relational. This was true even before morphology received an evolutionary interpretation. "Morphology is not the science of the fixed form, or Gestalt, but of the formation and transformations of organic form or Bildung" (Richards 1987: 151 n. 87). The term *morphology* covered all three aspects – study of form within one body, comparative anatomy, and embryology. First, the bodies of many plants and animals can be understood as generated by the repetition, with variation, of a sequence of similar elements (Goethe). Second, the bodies of taxonomically related organisms can be understood as variations on a type, with the type defined in terms of patterns of connections among corresponding elements (Geoffroy). These shared elements are more similar between organisms that are taxonomically closely related. They diverge as the taxonomic distance is greater. Third, adult bodies can be understood as the quasi-causal consequences of embryological development, and the patterns of embryological development again correlate with taxonomic relatedness (von Baer).

The latter two generalizations can be combined in a way that shows the richness of the morphological research program. Two species that are taxonomically distant from one another may have homological body parts that differ too greatly even to be recognized as homological. However, the early embryos are less differentiated than the adults, so it may be possible to trace the very different adult body parts back to their respective precursors in the developing embryo. Because the early embryonic precursors are less differentiated than the adult organs, any potential homological correspondence may be more apparent to the observer. Thus embryology provides a separate criterion, other than the principle of (adult) connections, for the homological identity of organs between distantly related species. The relative importance of embryology versus adult anatomy for the establishment of homology was a controversy within morphology throughout the century. The ambiguities that underlie that controversy still exist, and they are discussed in later chapters. Nevertheless, the parallel between the Natural System (as evidenced in comparative anatomy) and embryology allows a sort of triangulation of the concept of homology that demonstrates its robustness, even prior to the evolutionary interpretations of homological concepts.

That was morphology on the European continent around 1830. I must confess that I have underreported the extremism that could be found among the Naturphilosophen. That is not my concern. I am interested in the scientific content of the tradition. The idealism of continental morphology caused great disdain among British conservatives in the 1830s, and this attitude can still be detected among neo-Darwinians. I submit that this scientific program of morphology was idealistic in two possible senses, neither of them worthy of derogation. First, by the epistemological principles of the day, morphology involved abstraction to entities (homologies and archetypes) that are not directly observed by the naturalist. Such inferences were treated as idealizations at that time, whereas functional ascriptions were treated as empirically sound. Second, many of the naturalists involved in morphology were directly or indirectly influenced by Kant's transcendental idealism. Nevertheless, as discussed in Chapter 1, a Kantian viewpoint in itself is not consistent with many areas of modern thought.

Beginning with Geoffroy and von Baer, the concept of the morphological type persisted into the twentieth century. Synthesis Historiography emphasizes the ontological, idealist aspect of these types. "Ideal types" must exist in some ghostly neverland, or possibly in God's mind. I call this the *metaphysical concept* of type; it makes the reference to types look very foolish and antiscientific. In contrast, I will do my best to present the *explanatory concept* of type: Types were hypothesized in order to account for the wide and complex patterns of organic form, as those patterns have been discussed in this section. There is nothing foolish about them; some of the best thinkers of the nineteenth century were involved in their use. With a few exceptions (e.g., Agassiz), the metaphysical aspects of types were all but forgotten by the 1850s, whereas the explanatory aspects continued to be important throughout the century. It is impossible to understand the science of that period if we dismiss these theories as perniciously and metaphysically idealist.

3.3 NATURAL THEOLOGIANS ON UNITY OF TYPE

The first section of this chapter reviewed how biological adaptation became the basis of the AD. The second section reviewed the morphological program of Unity of Type as it developed in France and Germany. This section examines how British natural theologians reacted to Unity of Type. My purpose in this section is to assess the interpretation of morphology in Britain in the years before the *Origin* as an "idealist version of the Argument from Design" (IVAD, with my apologies for the ugly acronym). I argue that this interpretation distorts the true character of morphology in Britain. The IVAD interpretation would have been accepted by neither the natural theologians nor the British naturalists who accepted continental morphology. Even though the morphologists themselves varied greatly on species origins, I believe that only one important naturalist would have endorsed the IVAD. That was Louis Agassiz, and he did so in the late 1850s. The position is widely but falsely attributed to Richard Owen, an attribution that I discuss in Chapter 4. This section examines the attitudes of natural theologians and the early British advocates of morphology.

The purpose of an AD is to prove the existence of an intelligent creator of (fixed) species. The traditional, purposive AD did so by citing organic adaptations and claiming that only a mind could create them. Fixism was necessitated by the fact that the adaptations were designed for the species itself. "[I]f we allow . . . a *transmutation of species*, we abandon the belief in the adaptation of the structure of every creature to its destined mode of being" (Whewell 1837 v. 3: 574). As recent commentators tell the story, the IVAD served the same theological purpose as the AD, but it substituted idealism for adaptationism as its premise. Unity of Type was supposed to imply an idea in the mind of God, which presumably entails essentialism and species fixism. Let us examine the views of mainstream natural theologians regarding morphology. Did they conceive of Unity of Type as a theological equivalent to adaptation?

3.3.1 William Paley

Paley's *Natural Theology* was written before the advent of morphology. How-ever, Paley's discussion of the taxonomic unity of organic forms gives an im-portant hint as to natural theological reactions to Unity of Type. Paley cited taxonomic unity in order to prove only a single fact: the unity of the creator. He mentions organic unity three times, and he uses it each time to infer a single creator (rather than a committee; see Paley 1809: 66; 212; 449). The third mention, in Chapter XXV on "The Unity of the Deity," is particularly revealing. Here taxonomic unity is listed along with laws of astronomy and physics to prove the singularity of the creator. Recall that Paley describes astronomy as *"not* the best medium through which to prove the agency of an intelligent Creator"* (Paley 1809: 378; emphasis in the original). Paley clearly considers taxonomic unity as a secondary argument, used to establish the creator's attributes, not his existence. Secondary arguments lack the power to prove existence. Taxonomic unity merely proves the creator's singularity: The organic world was not a committee project.

3.3.2   William Buckland

Natural theologians of the *Bridgewater Treatise* generation differed immensely from Paley in the data they had available about life forms. Nevertheless, having stretched scientific minds to accept deep time and extinctions, the theologians of the Bridgewater generation were very protective of God's creation of individual species. Creation of species now had to take place successively through geological time, but it was special creation nonetheless. William Buckland was assigned the Bridgewater task of rationalizing the new geology with natural theology (Buckland 1836). Even though the fossils of extinct organisms were bizarre and unexpected, they fit into the taxonomic system designed for contemporary organisms. Buckland's use of the taxonomic categories differs only slightly from Paley's.

> [T]hese extinct forms of Organic Life were so closely allied, by Unity in the principles of their construction, to Classes, Orders, and Families, which make up the existing Animal and Vegetable Kingdoms, that they not only afford an argument of surpassing force, against the doctrines of the Atheist and the Polytheist; but supply a chain of connected evidence, amounting to demonstration, of the continuous Being, and of many of the highest Attributes of the One Living and True God. (Buckland 1836: viii)

The persistence of the taxonomic categories through extinctions and great reaches of time shows (1) that the intelligent creation of species didn't take place only once, but was a continuing process (refuting deism), and (2) that it was the work of only one creator (refuting polytheism). Unity does not prove intelligent creation per se, but only the singularity of the deity. In discussing the geological strata in which no fossils are found, Buckland falls back on the "prevalence of law." The law of the crystalline mineral ingredients of rocks also "attests the agency" of a mind (Buckland 1836: 46). But as always, nonpurposive law is acknowledged to be weak evidence. The "most obvious evidences of contrivance," and those that fill Buckland's *Treatise*, are the purposive functional details of the fossils themselves.

3.3.3   Charles Bell

Bell's *Bridgewater Treatise On the Hand* is by far the most biologically intriguing of the set. Bell was familiar not only with the Cuvier–Geoffroy debates but also with the German embryological studies. He accepts a system of correspondences within taxonomic categories, with no explicit comment on its basis. However, he flatly rejects the significance of any morphological research intended to extend those correspondences. It is a simple fact, but one not worthy of special study, that "the excellence of form now seen in the skeleton of man, was in the scheme of animal existence long previous to the formation of man" (Bell 1833: 22). The only topic worthy of study is the modifications of that scheme to suit the needs of the different animals in their different environments. Unity of Type, the theory behind "the more modern works" of morphology, is merely "a means of engaging us in very trifling pursuits – and of diverting the mind from the truth." The truth is that adaptation dominates the organic world (Bell 1833: 40).

Bell's skepticism about the value of morphology is not just a blanket condemnation. He cites specific cases. Bell gives special attention to a now-famous homology, that between the ossicles of the mammalian inner ear and jawbones of other classes of vertebrates. This homology replaced Geoffroy's alignment between ear bones and fish opercula, and it is still accepted today. But Bell treats the ear–jaw homology as a reductio ad absurdem of the entire program of Unity of Type.

> The only effect of this [ear/jaw] hypothesis is to make us lose sight of the principle which ought to direct us in the observation of such curious structures. . . . The first step ought to be to inquire into the fact, if there be any imperfection in the hearing of birds. That is easily answered – the hearing of birds is most acute. (Bell 1833: 137–138)

Bell reasons that *because* birds have perfect hearing, there can be no purpose in identifying one of their jawbones as corresponding to an ear bone in a mammal. In other words, the only conceivable reason to study morphology is to explain the failure of adaptation. Because adaptation never fails, morphology has no significance. This theme will recur in the twentieth-century arguments about adaptation and constraint. Anything worth studying is worth studying functionally. Unity of Type is a game of "trifling pursuits."

3.3.4   William Whewell

Whewell's Bridgewater assignment was astronomy and physics (Whewell 1836). The subject gave him little room to maneuver on biological topics. He did explain why teleology is inapplicable to researches in the physical sciences (and so why his treatise was harder to write than the others). He reinterprets Bacon's aphorism that likens final causes to vestal virgins (pious but barren) as saying that final causes should not be used *in the discovery of physical laws*. However, after one has discovered a system of physical laws without assuming final causes, one can look back at them and assess how beneficial they are in total.

Whewell had a better opportunity to discuss purposive biology a year later in his *History of the Inductive Sciences*. Like Buckland, Bell, and others, Whewell gives a long and lavish description of the superior empirical basis of Cuvier's functionalist approach, and its implications of species fixism. Given Whewell's Kantian leaning, we might expect him more to be sympathetic to idealist biology than his empiricist colleagues were. He had no such sympathies. He soundly rejects Geoffroy's nonpurposive program. He singles out for special criticism Geoffroy's statement that he "take[s] care not to ascribe to God any intention" (Whewell 1837 v. 3: 461). He cites Kant's approval of the teleological study of organisms, and he asserts that teleology is a necessary assumption of any study of organisms. Whewell's *Bridgewater Treatise* had stated that final causes must *never* be used in physical research; his *History* states that final causes must *always* be used in biological research. Unity of Type illegitimately ignores the purposes of the organs under study.

Twenty years later in the third edition of the work, Whewell would modify these views and grudgingly accept homologies (for external reasons discussed in Ruse 1979: 149). Whewell's reluctance is obvious. Even Owen's pious version of Unity of Type is said to be "a view quite different from that which is described by speaking of 'Final Causes,' and one much more difficult to present in a lucid manner to ordinary minds" (Whewell 1863: 644). Compare this with Paley on astronomy and Buckland on crystallography. Nonpurposive laws are not useful to natural theologians, and they admit it.

3.3.5   Peter Mark Roget

Roget's *Bridgewater Treatise* is a striking exception to the otherwise universal rejection of Unity of Type by natural theologians of the 1830s (Roget 1834). Most of the work is an uncontroversial report on functional anatomy. The final chapter, entitled "Unity of Design," turns to idealist biology. Roget calmly, pleasantly, and somewhat superficially reports on a very wide selection of the results of idealist biology. He gives no hint of controversy. Six pages from the conclusion, just after a discussion of the Meckel–Serres law of parallelism, Roget seems to realize that it is time to change his tone. He assures the reader that species really are fixed after all, mildly criticizes the extravagance of Lamarck and Serres, praises Newton for his humility, and ends with a few pious words of thanksgiving and praise. This is the earliest discussion of idealist biology by a natural theologian, and it endorses both idealist biology and natural theology. Is his book a serious argument that idealist biology entails special creationism? Let us consider some of the details.

Whether Roget is expressing the IVAD, idealist natural theology, depends on what he infers from idealist biology. Does Unity of Type, by itself and independent of adaptation, imply the existence of God and the fixity of species? No. Roget's brief statement about species fixity is reported as a scientific fact (Roget 1834 v. 2: 636), and it is attributed not to an idealist morphologist, but to Cuvier! If it doesn't prove species fixism, then what does idealist biology prove about the existence of God? Roget states only one implication, the same one cited by Paley and Buckland. Unity of Type refutes polytheism by proving that all species have all "emanated from the same Creator" (Roget 1834 v.1: 52).

So Roget's discussion of idealist biology gives no grounds for believing in God's existence. Those grounds come only from the ordinary adaptationist AD as it is repeated throughout the treatise. Idealism gives no grounds for believing in species fixism. Those grounds come only from Cuvier, who was not an idealist but an adaptationist. Just as Whewell was unable to prove God's existence from the marvelous details of the vast reaches of astronomy, Roget is unable to do it from idealist biology. The facts of idealist morphology, like the facts of astronomy, show you how mysterious and marvelous the world is. They are illustrations of God's power, not proofs of his existence. They are certainly not proofs of species fixism.[18]

Roget's *Bridgewater Treatise* is the only good case for the IVAD prior to Richard Owen's work from 1846 to 1849. A close reading shows that it was not an idealist version of the AD at all. The empirical details of unity had changed since Paley, but the theological significance had not. Its only value was as evidence for monotheism over polytheism. The proof of an intelligent creator depended entirely on the adaptationist AD. The next chapter is devoted to Richard Owen and his relation to Darwin. For now, let us examine the way in which continental biology was actually imported into British thought.

3.4   THE STRUCTURAL TURN

Even though Cuvier's status in the scientific community was high after the disputes of 1830, the functionalist orientation that he shared with the British

---

[18] If he is not constructing an IVAD, then why does Roget bring idealist biology into a *Bridgewater Treatise*? Ironically, he seems to have been trying to defuse the radical implications that many contemporaries were deriving from the study (Desmond 1989: 229–230).

natural theologians eroded during that decade. Younger naturalists were attracted to structuralist studies of embryology and comparative anatomy. Medical radicals had spread the awareness of the continental theories in Britain (Desmond 1989). A more conservative source of interest flowed through the romantic poet Coleridge and his associates, eventually expressing itself in the work of Richard Owen (Sloan 1992). Roget's pious Bridgewater discussion may have reduced the stigma of radicalism, but it had little effect on the religious discussion. A short spurt in popularity of idealism among theologians would wait until Owen's work. Nevertheless, British scientific interest in idealist continental theories gradually grew.

This interest can be seen in the work of two young scholars who graduated from Edinburgh medical school. Martin Barry matriculated in 1833 and had studied on the continent both before and after his degree. William Carpenter arrived in 1835 after beginning his medical studies at the University College London, where he attended lectures of the radical Robert Edmond Grant (Rehbock 1985: 59–63). Barry and Carpenter are indicative of the trend toward morphology and away from adaptation during the 1830s.

3.4.1 Martin Barry

Barry's two-part "On the Unity of Structure in the Animal Kingdom" (Barry 1837a, 1837b) was the first published report in English of von Baer's embryological doctrines. It is far more thorough and serious than Roget's chatty treatise, and in my opinion one of the gems of early British structuralism. According to Barry, the Unity of Type among all animals derives from the fact that they start in their development from the same point ("*germs* from Infusoria to Man, are essentially the same") and develop in the "same manner," from homogeneous to heterogeneous conditions. Divergence among species is generated by the fact that organisms differ in the "*direction*, or *type*" (but not the manner) of their development. "And are we not led fairly to the conclusion, *that all the varieties of structure in the animal kingdom, are but modifications of, essentially, one and the same fundamental form?*" (Barry 1837a: 126–127; emphasis in the original)

Barry clearly reports on the embryological grounds for the impossibility of transmutations between classes (i.e., phyla). The characters that mark classes are the earliest to differentiate in the embryo. These characters set the direction of development and serve as necessary causal conditions for further development. For this reason "it is absurd to say, that one Class of animals can pass into another; such, for example, as the Cephalopoda of the class Molluska, or the Crustacea of the class Articulata, into Fishes of the Class Vertebrata"

(Barry 1837a: 130–131n.). Barry here rejects the Meckel–Serres law of parallelism that had been made endorsed by the Geoffroyans (and reported favorably by Roget). Barry's concern is *not* to deny transmutation of species. That topic is not even under discussion. Rather he is showing an implication of von Baer's embryology; the Natural System (as evidenced by embryology) is treelike rather than linear. Barry even offers a sort of developmental theory of heredity to explain how adult forms are derived, by means of development, from the single germ.[19]

The correlation between embryological development and the taxonomic categories of the Natural System is extremely significant. It implies that embryology is a better guide to classification than adult structure, where (Barry says) "function tends to embarrass." That is, the adaptations of common adult structural elements to diverse purposes tend to mislead us when we try to establish accurate affinities. Attention to embryology can help us avoid these embarrassments. Like Strickland in Chapter 2, Barry is a fallibilist about classification. Our current categories may be wrong, and we should always be ready to improve them. Recall that Strickland had also rejected the possibility that "transitional" species showed real affinities to more than one large group. Strickland's refusal was based on his faith in the reality of the Natural System, and its determinate structure of groups within groups. Barry has a more deeply causal reason to reject intermediate forms. Bodies of organisms begin as embryos, and these develop by a process of increasing differentiation, during which the characters of classes, then of orders, then of families, and so on become apparent. This generalization is well established by microscopic observation. The fact that adult hagfish appear to be intermediate between fish and mollusks is a mere "embarrassment," presumably caused by adaptation. A careful embryological study of the organism would determine which group it really belongs in.

Barry does not say, as Strickland does, that *all mollusks* are equally distinct from *all vertebrates*, and *all birds* from *all mammals*. However, he has even better reason than Strickland to make that claim, because he has a causal explanation of the branching structure of the Natural System. The branching Natural System is the outcome of the branching of embryological development. *Of course* adult organisms can be classified in a hierarchical manner. Their bodies are built by causal processes that are identical at their beginning stages, and that gradually diverge as they proceed.

In the second half of the paper, Barry illustrates his rejection of the Meckel–Serres parallelism with a remarkable diagram captioned "The Tree of Animal

---

[19] Barry's theory of heredity is discussed in Chapter 7.

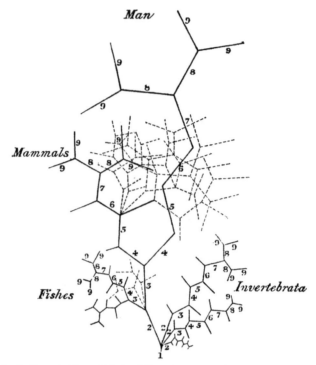

Fig. 1. Martin Barry's Tree of Animal Development (1837b: 346), representing von Baer's laws. Although Barry did not commit himself to evolution, he believed that an individual embryo's path through the branches of the tree was controlled by its heredity, with distinct hereditary "properties" expressed in the various segments of its path (see Chapter 7, this volume).

Development," which shows a branching structure spread through three-dimensional space (Barry 1837b: 346). It is impossible for a modern reader to resist seeing this as a phylogenetic tree (see Figure 1). It is intended to depict the branching course of embryological development among animal species. At the end of the paper he again stresses the importance of embryology for classification. The attention of nonembryological systematists has been "directed to the grouping of the *twigs* – as if thus they were to find their natural connections, without even looking for assistance to the branches, or the trunk that gave them forth" (Barry 1837b: 362).

3.4.2   William Carpenter

Carpenter's view on structuralism was expressed in his 1838 review of Whewell's *History of the Inductive Sciences*. Carpenter compared Whewell's

History to his *Treatise* of the previous year. The *Treatise* had claimed that physical research must *avoid* final causes. The *History* claimed that morphological research must *assume* final causes. Carpenter rejected the double standard. If objective knowledge of physics requires a suspension of belief in final causes, why should objective knowledge in physiology require the opposite (Carpenter 1838: 338)?[20] Carpenter recognizes that final causation can be useful in discovery of low-level facts in physiology, and he discusses the human skeletal–muscular system as an example.

> The teleologist would say, with apparent truth, that each of the bony processes was intended for the attachment of a muscle. . . . On the other hand, the philosophical anatomist, fully acknowledging the adaptation between the osseous and muscular systems, would disregard it for the time, whilst seeking for the laws regulating the development of those systems. . . . Thus, he would find that each of the important processes in the human skeleton exists as a separate bone in some of the inferior animals. (Carpenter 1838: 340)

Carpenter recognized that Whewell's adaptationist reasoning enabled his taxonomic nominalism; biological functions do not require structuralist comparisons. However, Carpenter valued the "higher laws" of Unity of Type, which come from nonteleological reasoning. These can *causally* explain the adjustments of muscle to bone that occur during embryonic development in all vertebrate species.

Unlike Barry's work, Carpenter's review of Whewell was concerned with religion. Like other liberal theologians, Carpenter considered the higher laws to "imply a higher degree of Creative Wisdom and Power . . . than that the formation and adaptation of each separate muscle and each individual process required a distinct effort of creative skill" (Carpenter 1838: 340). Carpenter argues for an almost deist commitment to exceptionless universal laws. Such laws were not to be discovered by purposive reasoning. As we will see, Carpenter was one of a number of liberal theologians to consider the possibility of naturalist species origins. He later reported that he "had not the least objection, either philosophical or theological, to the doctrine of Progressive Development, if only it could be shown to have a really scientific basis" (Carpenter 1889: 108). Whewell's insistence on a double standard for physical and biological science reflected his steadfast opposition to the notion.

---

[20] In defense of Whewell, Kant had shown a very good reason why teleology was involved in morphology, at least when embryology was taken into account. Whewell was quite aware of Kant's arguments, although he seems to have been one of the few British thinkers to be concerned with them. Neither Carpenter nor Huxley nor even Darwin regarded von Baer's Kantian teleomechanism to be a barrier to their theories.

Carpenter favored idealist structuralism over teleological adaptationism, but he did not give an AD in anything like the traditional sense. Far from citing morphology in support of special creationism, Carpenter virtually endorses a Christian evolutionism.

3.4.3 Rudolf Leuckart

The work of Leuckart is quite distinct in context from that of Barry and Carpenter. He was a German morphologist whose important work began in the 1840s. The scientific reactions to his views illustrate an important aspect of the structural turn. In the 1840s, the phenomenon of *alternation of generations* was beginning to be understood. Alternation of generations can be seen as an elaboration of metamorphic life cycles. Organisms that metamorphose, for example a caterpillar to a butterfly, change their morphology from a larval to an adult form. The larval form feeds and grows but does not reproduce. Alternation of generations occurs when a species can take on two or more distinct morphologies, but more than one form can reproduce. Typically the simpler "larval" stage reproduces asexually, and the more complex "adult" stage reproduces sexually. The alternation had originally been discovered in jellyfish and polyps, playing havoc with attempts at classification (Winsor 1976: Ch. 3). Leuckart proposed an adaptive explanation of the alternation of generations. He argued that the production of large numbers of offspring is clearly beneficial to the survival of species, and so the asexual reproduction of larval forms was merely a way to increase the effective number of offspring. For organisms that had high mortality rates of larvae, alternation of generations would be beneficial to the survival of the species.

This explanation makes perfect sense to adaptationists, whether Darwinian or Paleyan. It answers Plato's question, "Why is it good?" But Leuckart was writing in the late 1840s. That kind of reasoning was no longer acceptable within science. His explanation was criticized as circular reasoning (Winsor 1976: 72). Proper scientific method did not allow one to dream up benefits of a phenomenon and then claim to have *explained* that phenomenon. Adaptationism was so out of favor that even Leuckart's reasonable-sounding adaptationist hypothesis was rejected as unscientific.

Note the contrast between Geoffroy's day and Leuckart's. In the 1820s, Geoffroy was forced to confess to so-called idealism because he dealt in nonfunctional correspondences. After the structural turn, functional correspondences were regarded with skepticism; Leuckart was criticized for his adaptationism reasoning. Standards change, and change again. The good news for Leuckart is that he survived long enough for Darwin to relegitimize his adaptationist leanings. In 1869, Leuckart was again able to say, without fear of criticism, that a major task of science was "to show how the specific conditions of adaptation" affected "the forms of living individuals" (Nyhart 1995: 181). Adaptation was no longer a mere consequence of form, but a cause of it – just as it had been for Cuvier fifty years earlier.

Darwin convinced the world that natural selection existed, but not that it was the primary cause of species change. One reason was that, in Darwin's day, adaptation was no longer a central target of scientific explanation.[21] It is very hard to appreciate this fact from our modern perspective. Neo-Darwinians share with Cuvier, Paley, and the *Bridgewater Treatise* authors the evaluative judgment that *the most important fact of biology is adaptation*. Most biologists of the 1850s did not share that judgment. Adaptationist biology was old-fashioned; structural biology set the agenda. Idealistic morphology supplanted natural theology; it did not contribute to it.

3.5 WHAT IS NATURAL THEOLOGY?

In this chapter I introduced the study of idealistic morphology and examined its relations with natural theology. I submit that Unity of Type did not take the place of adaptation within a modified AD, at least for the authors discussed in this chapter. Unity of Type was always seen to be a challenge to natural theology. The growth of interest in morphology in Britain constituted a loss of interest in adaptation and a deterioration of the scientific significance of natural theology. The AD was an argument for species fixism, an essentially antievolutionary view. Almost all of the modern authors who depict idealistic morphology as an IVAD do so because they believe that idealism was antievolutionary.[22] If idealism was merely a variant of natural theology, then it was an obstacle to evolution. Evidence is to the contrary. Morphology was not an IVAD, and it was important to the development of evolutionary thought.

The reader may feel that I am biasing the case against the IVAD by defining the AD so narrowly that an idealist version could never exist – so let us discuss the AD in more detail. In Section 3.1 I distinguished two types of natural

---

[21] Other reasons are discussed in what follows.

[22] I have neglected a complication in Bowler's 1977 paper. He actually argues that the "idealistic version" of the AD should be seen as anti-Darwinian, not just antievolutionary. For that reason, he attributes to IVAD many of the "non-Darwinian" evolutionary theories of the late nineteenth century. Because Bowler's views on the late nineteenth century are discussed at length in Chapter 6, I do not address them here. Everyone except Bowler uses the IVAD to explain resistance to a belief in evolutionary change.

theological arguments, primary and secondary arguments. The AD of the nineteenth century was a primary argument establishing an intelligent creator on the basis of empirical observations of special adaptations. The fact that the adaptations were special – that is, directed toward the benefits of individual species and their members – was crucial. It proved that individual species had been the entities that were intelligently created. As we saw in Section 3.3, the AD opposed evolution because of this belief in special adaptations. For this reason, if a natural theological argument does not entail special creationism, it is not a primary argument; that is, it is not an AD in the meaning of the term. If an idealistic version of the AD really was a primary argument for intelligent design, then it must have been taken to imply species fixism. But why should idealism (as we have seen it) imply species fixism? Perhaps it would, if idealism were conceptually tied to Platonic essentialism and essentialism in turn to species fixism. That is, the SH principle of the IVAD might be supported by the SH Essentialism Story. In fact, though, *no one during this period treated idealism as grounds for species fixism*. There is no historical evidence either for the Essentialism Story or for the IVAD. The fact that one might support the other is irrelevant.

There is one more way that the IVAD might be saved. In my account, the AD must imply species fixism. Perhaps this assumes a too-narrow view of natural theology. Like science itself, natural religion changes through time. Couldn't a broadminded natural theologian regard naturalistic species origins as merely the means used by God to achieve the ends of creation? Couldn't idealism be used as evidence of the plan by which the creation was achieved?

Let us call this view *liberal natural theology*. In just this spirit, Robert Chambers anonymously published *Vestiges of the Natural History of Creation* in 1844. He argued for a form of progressive evolution, basing the argument on popularized versions of dozens of scientific and marginally scientific theories of the day, including idealist biology. Paley and the *Bridgewater Treatise* authors are treated with respect, and they are said to have proven "that [adaptive] *design* presided in the creation of the whole [world] – design again implying a designer, another word for a *Creator*" (Chambers 1844: 234; emphasis in the original). Chambers sees himself as a liberal natural theologian in our sense. Should we consider the *Vestiges* as a work of liberal natural theology, and thereby see idealist morphology as an IVAD? This seems to be only a decision about labels. If we broaden our view of natural theology to include evolutionary theories, then idealist biology can certainly be included in it.

There is one problem in this expansion. If natural theology were broadened to include Chambers and other liberals such as Carpenter and Baden Powell, it would have to include people whom we do *not* think of as

74

advocates of natural religion. In particular, it would have to include Charles Darwin himself! Darwin personally paid for the publication of a pamphlet entitled *Natural Selection Not Inconsistent with Natural Theology*, authored by Asa Gray. He sent the pamphlet to naturalists and divines, and he even published an advertisement for the pamphlet in the 1861 third edition of the *Origin* (Moore 1991: 369). I assume that the SH authors who support the IVAD do not wish to depict Charles Darwin as one of its advocates.

In conclusion, if the AD is conceived as a challenge to evolution, there is no justification for the claim that an IVAD existed during the time period so far discussed. To include idealistic biology within natural theology, we would need to broaden our concept of natural theology. As soon as we do that, natural theology itself includes evolution. However, there is one important figure whose writings might falsify these historical conclusions. It is the allegedly Platonic morphologist Richard Owen. His work and its relation to Darwin are important enough to deserve a separate chapter.

4

Owen and Darwin, the Archetype
and the Ancestor

4.1 INTRODUCTION

Richard Owen was the most prominent and respected British naturalist of the 1840s and 1850s, and he was an active researcher for nearly sixty years. He was the transcendental anatomist who was closest to Darwin while Darwin's theory was being formulated. For this reason, he was important to Part 2 of the *Origin*, the argument for the fact of common descent. Owen's reputation was first established as a Cuvierian functionalist, a mode of thought that coincided nicely with the Bridgewater generation of natural theology. During the 1840s he followed the structural turn of many British naturalists toward continental morphology (Chapter 3, Section 3.4). He gathered, organized, and continued the work of idealist morphologists. He presented it to the public in an orderly and a quite empiricist-sounding style. He clarified the concepts of homology and analogy, he revised and rationalized the system of the naming of bones of the vertebrate skeleton, he articulated three distinct kinds of homology, and out of these he constructed the Vertebrate Archetype. He acknowledged the work of the continental morphologists, but he was careful to distance himself from their perceived excesses. In this way he was able to present his own version of the morphological results as the result of careful induction, and not idealist speculation. The least recognized of Owen's achievements (especially among modern commentators) is that he associated Unity of Type and the Vertebrate Archetype with a naturalistic cause of the origin of species.

Owen had ties to two distinct groups of patrons, and it was important to his career that he maintained them. One was the traditional Oxbridge establishment, consisting largely of conservative Bridgewater thinkers. The other was a London circle of intellectuals who surrounded the transcendental poet Samuel Coleridge. This group was less antagonistic to continental thought, with its idealist and pantheist overtones. Owen's own structural turn was probably encouraged by his London associations, but (as we will see) it was resisted by the Oxbridge conservatives.

Owen tested the tolerance of his conservative patrons with a research program that had quite radical implications. He knew that the conservatives would vet his ideas, so he hedged his statements carefully. He couched his most radical ideas in pious terms. The Bridgewater generation had used the same technique when they discussed the recently heretical facts of geological time, extinction, and successive species origins. Their piety in reporting the new geological discoveries, together with their staunch species fixism, had satisfied all but the most diligent biblical literalists.[23] Owen's similar use of piety to camouflage radical ideas had paradoxical results. The conservatives saw *through* the piety, and they condemned his radicalism. Later, the Darwinian liberals saw *only* the piety, and they failed to recognize Owen's radical ideas at all.

The Darwinian liberals accepted Owen's technical achievements, such as the Vertebrate Archetype and his clear definitions of homology and analogy. They derived their own radical results from them. Then they condemned Owen for his conservatism. In the first edition of the *Origin*, Darwin identified Owen as a species fixist, an assertion that he retracted in the second edition. Huxley developed a personal and professional antagonism toward Owen from the mid-1850s onward, and this may have contributed to Darwin's gradual estrangement from Owen. The Darwinians (including Huxley) took control of the scientific establishment in the later part of the century, and Huxley was (amazingly) invited by Owen's grandson to compose a memorial to him (Huxley 1894). This settled Owen's historical reputation, and in 1960 he was described as a "now forgotten naturalist" (Gillespie 1960: 313). In modern popular writing he is still often described as a special creationist and antievolutionist, and many of the Darwinians' distortions of his views are treated as factual history.[24] He was damned as a transmutationist by the creationists, and as a creationist by the transmutationists. The transmutationists eventually won the day, and Owen's reputation as an antievolutionist was the one that passed on to posterity. I must admit that even sympathetic reports on Owen's personality do not make him an appealing figure. However, our concerns are neither with Owen's reported haughtiness nor Darwin's reported modesty.

---

[23] By the early nineteenth century, the biblical literalists were of the opinion that species fixism was literally stated in Genesis. This had not been the opinion of the literalists of the seventeenth century (Zirkle 1959). This raises obvious questions about what *literalism* means.

[24] Owen's biographer, Nikolaas Rupke, details several of these distortions, including the often-repeated but mistaken report that Owen had claimed priority on the concept of natural selection itself (Rupke 1994: 246 ff.).

They are with the scientific contribution of Owen's ideas to the subsequent development of evolutionary biology.

Typology is an important topic in our study for two reasons: one positive and one negative. On the positive side, it was a central theoretical concept of nineteenth-century morphology. On the negative side, "typological think-ing" was fingered as the very antithesis of scientific evolutionary thought in 1959, and that stigma forms an important aspect of the Essentialism Story. To understand the history of structuralist biology and its contributions to evolu-tionary thought, we must come to terms with the condemnation of typological thinking that has been so widespread since 1959.

Chapter 3 introduced morphology and the morphological type. I distin-guished between the metaphysical concept of type (according to which mor-phological types are seen as idealist dogma) and the explanatory concept of type (according to which morphological types played a role in explaining organic form). The notion that morphological types were metaphysically as-sociated with ideas in God's mind seems to have been behind the notion of the idealist version of the argument from design that was criticized in Chapter 3. In actual fact, the individual morphologists had a variety of ways of interpret-ing morphological type, and these became less metaphysical as time passed (Nyhart 1995). Paul Farber has reviewed the various concepts of type that were in play during the early nineteenth century: the *collection-type concept,* the *classification-type concept,* and the *morphological-type concept* (Farber 1976). The collection-type concept refers to the museum practice that desig-nates one individual specimen as the *type specimen* of a species. The practice may have indicated an unjustified faith in limited variation within species, but it had little theoretical impact. The classification-type concept was referred to as the exemplary type in Chapter 2 (following Winsor 2003). Far from es-sentialist, the exemplary type was especially designed to allow for polythetic taxonomic groups. From the standpoint of metaphysics, the morphological type is where the action is. The morphological type is an abstraction, whereas the exemplary and taxonomic types are tangible entities. If *typology* carries the antievolutionary implications that are alleged of it, the morphological type must be the culprit.

Our search for essentialist reasoning in Chapter 2 bore no fruit prior to the 1840s. Then we saw Strickland appealed to "essential characters" as those that revealed the genuine affinities between species and groups. Soon we will see how Owen claimed that the relation of homology reflects the "essential nature" of animal body parts. Unlike collection types and classification types, morphological types were invoked in causal (or quasi-causal) laws. This means, in philosophy-speak, that they justify predictions and support counterfactuals. In more ordinary language, they describe how things *must be* rather than merely how they have been observed to be. The exemplary type merely summarized what characters *happened to be* shared within a group. The morphological type explains *why* the characters are shared as they are. Mammals have fewer jawbones than birds *because* they have more ear ossicles. Vertebrates all have single proximal forelimb bones (the humerus) *because* the Vertebrate Archetype has single proximal forelimb bones; they all have single proximal hindlimb bones *because* the vertebrate forelimb is serially homologous ("the same" in the serial sense) to the hindlimb. Morphological types are nomological; they refer to the lawlike structure of biological reality. This differentiates them from Cuvierian nominalist taxonomy.

The twentieth-century discussion of morphological types has been strongly conditioned by Mayr's 1959 identification of typological thinking as the foe of evolutionary thought throughout the ages. Mayr's publications of this era were supported by his reading of only one set of primary sources, those of Louis Agassiz, Mayr's predecessor as Director of the Museum of Comparative Zoology at Harvard. His historical reports were greatly "indebted" (Mayr's term) to Lovejoy's *Great Chain of Being* (Lovejoy 1936; Mayr 1976: 254 n. 1). Lovejoy's book dwells much more on the scala natura and its associated continuity than on essentialism and discontinuity. Mayr was interested only in the discontinuity. He was correct in one way about Agassiz – he was a species fixist, and he argued for fixism on the grounds of the eternal changelessness of ideas. It is questionable whether these opinions are traceable to his idealist teachers, as Mayr alleges (Winsor 1979). It is even more questionable that Agassiz's combination of idealism and species fixism was shared with other important nineteenth-century figures, Mayr's allegations to the contrary. Although Mayr eventually became extremely well read in the history of his discipline, his great generalization of 1959 regarding typological thinking and species fixism in the history of Western thought was based on very slim empirical evidence. Nevertheless, the nineteenth-century morphologists *were* typologists, and when historians and philosophers began to read the morphologists, Mayr's generalization rang in their ears: They read the typology as a form of antievolutionism.

William Coleman's 1976 "Morphology Between Type Concept and Descent Theory" is an important early paper on nineteenth-century morphology. It treats type and descent as exclusive categories; one cannot believe in both. Haeckel and Gegenbaur are said to have abandoned the type concept in favor of descent theory. Morphological types are a prioristic presuppositions that *governed* the study of diversity, rather than discoveries *inferred from* the study of the organisms themselves. "The morphologist sought to circumscribe these types and distribute known animal forms among them" (Coleman 1976: 150). Notice the similarity to the principle of the empirical accessibility of function: Structural similarities are based on a priori reasoning.

As historians delved deeper into the period, the contrast between descent and typology became less exclusive. By the 1990s, most commentators had come to treat the morphological type as a theoretical device that was used both in preevolutionary and evolutionary frameworks (Di Gregorio 1995; Nyhart 1995; Bowler 1996; Lyons 1999; Camardi 2001). Thomas Henry Huxley, after all, was a student of morphological types both before and after his conversion to Darwinism. The type concept served as a point of organization around which morphological inquiry proceeded. The continuity between the pre- and post-Darwinian concepts of type is important to recognize (contrary to Coleman). This continuity illustrates the distinctly morphological scientific interest in *explaining form* (Nyhart 1995).[25]

The recognition that typology was evolutionary during the late nineteenth century leaves open the question whether it had been antievolutionary prior to Darwin. Why was it thought to be antievolutionary? The answer is clear from Mayr's earliest writings on typological thinking. Types are timeless entities, for which change is impossible. Nevertheless, timelessness of types still does not imply species fixism *unless* species themselves are types. To my knowledge, that question has never even been addressed. So the crucial question is this: *Were species thought to be types?*

The first named "types" in the morphological literature were the four phyla that correspond to Cuvier's embranchements: Articulata, Vertebrata, Molluska, and Radiata. Lower-level taxa were soon identified with their own types. The mollusk type had within it a cephalopod type and a bivalve type; the vertebrate type had types for fish, birds, mammals, and reptiles. Did the

---

[25] Though these historians did not themselves contrast their work with the earlier discussions, the difference is apparent. I take it to indicate the development of an autonomous tradition in the history of science that is more independent of contemporary scientific interests than are the interests of Mayr and most philosophers of science (myself included).

types extend all the way down to the species level? I believe that the answer is clear on reading the literature.

Species are not types.

Never were, never could be. Morphological types were hypothesized as a way of representing the patterns of unity *between, not within* species. Even Louis Agassiz, the only genuinely Platonic species fixist, did not treat individual species as types.

> [Agassiz] and indeed most of his contemporaries discussed the anatomy, embryology, or other characteristics of species only as the species were representatives of their order or class, because questions of theoretical interest concerned the larger groups. (Winsor 1976: 132)

When the morphological type was invented, there was no reason to puzzle over the unity that exists *within* a species. Species were generally assumed to be limited in their variation, and had been so assumed since the time of Linnaeus (Chapter 2, Section 2.2). As Mayr and others have pointed out, the study of within-species variation was of very little interest before Darwin. In this sense, Mayr is exactly right that population thinking was a major innovation. But he is wrong (I believe) in his claim that the only alternative to population thinking was Platonism about species. Another alternative, neither population-thinking nor Platonic typology, is to think of species as *contingently* fixed, as breeding true not because of a metaphysical dictum from Plato but because of the facts of how organic reproduction works.[26]

I consider morphological typology to have been a necessary step toward evolutionary thinking because it contributed to the development of taxonomic realism, that is, to the establishment of the Natural System as a real objective structure in nature. Species fixism was most easily defended by taxonomic nominalists, such as Cuvier and the Bridgewater authors. They denied real groups and considered the Natural System to be merely a convenient information-storage device. Of course it was *possible* to be both a typologist and a species fixist, but, in actual fact, typology was never used as grounds for species fixism. The development of the morphological type, along with its hierarchical structure of subtypes and its associations with embryological development and the fossil record, was a major contribution to this increased

---

[26] Coleman's assertion that typology implies species fixism originated in the 1960s, soon after Mayr's influential paper (Coleman 1964: 102, 146). However, Peter Bowler has perpetuated it, and sometimes even defines typology in terms of species *but not* higher taxa (Bowler 1984: 101; Bowler 1999).

realism about the Natural System, and so ultimately to an evolutionary view of life. This can be seen in the work of Richard Owen, and in Charles Darwin's use of that work.

4.3 OWEN BUILDS THE ARCHETYPE

Richard Owen's achievement was to strengthen and articulate the morphological type concept and mold it into the Vertebrate Archetype. He did this in an atmosphere of theological tension and conflicting social and scientific class interests. Prior to Owen's work, morphological types had been associated with two strains of continental thought that were highly disapproved of in Britain, one epistemological and the other theological. The British scientific establishment was distrustful of the speculative and nonempiricist aspects of continental thought. The theological establishment was wary of the pantheism that seemed to accompany such thought. Owen had to calm both fears while retaining what he considered the scientifically important aspects of the morphological type concept. To be sure, he was not a cutting-edge revolutionary even in his archetype work. Edinburgh-influenced naturalists such as Knox, Grant, Barry, and Carpenter had preceded Owen in the English-language expressions of structuralist continental ideas (Desmond 1989). Owen had at first been "the British Cuvier," the predominant conservative naturalist. During the 1830s he followed Cuvier's functionalist program, and he fit nicely within the Bridgewater world (Rupke 1994: 117). Owen's archetype work in the 1840s was to belie his Cuvierian reputation and eventually to swing the mainstream of British natural history into the structuralist camp.

To accomplish this, Owen had to calm the fears of both the empiricists and the theists. One tactic was to endorse a conciliatory position between Unity of Type and adaptation. This was mere rhetoric: "Owen presented his position as a halfway point between the pure teleology of Charles Bell and the pure morphology of Geoffroy, but in fact the distinction has meaning only for the morphologist" (Ospovat 1981: 130). A more substantive (but still rhetorical) achievement was an improvement in the epistemological packaging of the idealist theories, combined with a creative massaging of the Bridgewater theological intuitions. However, Owen's real scientific achievement was to catalog and rationalize an immense amount of continental vertebrate anatomy and to devise a theoretical framework that could present it in a unified form, the Vertebrate Archetype.

This occurred in three steps. First, Owen clearly distinguished between *analogy* and *homology*:

Analogue. – A part or organ in one animal which has the same function as
another part or organ in a different animal. . . .
Homologue. – The same organ in different animals under every variety of form
and function. (Owen 1843: 374, 379)

A distinction similar to this one had been implicit in earlier writings of continental morphologists and others. One term (sometimes *affinity*, sometimes
homology) was taken to designate the deep and meaningful similarities between organisms – those that revealed the underlying Unity of Type. A contrasting term indicated superficial similarities. Darwin was unfamiliar with
Owen's 1843 distinction when he composed his unpublished *Essay* of 1844,
which referred to the "ill-defined distinction between true and adaptive affinities" (Darwin 1909: 215). Owen's definition standardized the terminology,
and it stipulated that superficial resemblances were due to functional similarities. Owen's structuralism is revealed even at this early stage of the discussion.
Homology is based on structure (not function), and it provides the deepest
insights into organic nature. Darwin had been skeptical about "true affinities"
in his Essay. However, after recognizing Owen's clarification, Darwin had no
hesitation in using homology extensively throughout the *Origin*.

Owen's second step was to document the fact that the *entire skeletons* of
vertebrate groups could be shown to correspond, bone for bone, with other vertebrate groups. Previous morphologists had mostly been content to search for
surprising correspondences, such as the homologs of mammalian ear ossicles
in nonmammal groups. In *The Archetype and Homologies of the Vertebrate
Skeleton*, Owen cataloged the various names by which vertebrate bones had
been designated by the specialist–anatomists who had named them (Owen
1848). Owen had been a member of Strickland's committee on nomenclature,
and he had gained from the experience. The 1842 BAAS statement on nomenclature had shown how a good British empiricist (John Locke in fact) chose
names. Names were arbitrary symbols; they should not be loaded down with
theories or descriptions, because theory-laden names would be a burden on
the future progress of science (see Chapter 2, Section 2.5). Bird anatomists,
fish anatomists, horse anatomists, and human anatomists had used different
names of bones, often reflecting a superficial resemblance that the bone possessed in their group. Owen tabulated the various names and descriptions by
which all of the bones were known, usually separately for different groups
(fish, horses, humans, etc.). He named each, often replacing a long anatomical
description with a brief name.

Owen described the renaming project as if it were a simple empirical
catalog with no theoretical ambitions or presumptions. This was a strategic

posture, made necessary by the Bridgewater-era principle of the empirical accessibility of function. Recall that Geoffroy had been labeled as an idealist because of his recognition of structural correspondences. Owen wanted to circumvent that burden. To do so, he disguises one immense theoretical assertion as a simple empirical fact. The slight-of-hand appears in this passage:

> To substitute names [of bones] for phrases is not only allowable, but I believe it to be indispensable to the right progress of anatomy; but such names must be arbitrary, or at least, *should have no other signification than the homological one.* (Owen 1848: 3; emphasis added)

The underlying theoretical assumption of the entire project is that all vertebrates are built on a single body plan. Homologies are the elements of this body plan. This is a direct contradiction of the classificatory doctrines of Cuvier and the Bridgewater natural theologians. It is done in such a brisk, no-nonsense manner that the radicalness of the project was unnoticed. The practice that Bell had condemned as "trivial pursuits," and that others suspected of various sins from idealism to pantheism to atheism, was repackaged by Owen into an apparently harmless form.

Owen's third step was to articulate three distinct kinds of homology: serial homology, special homology, and general homology. Serial homology was the relation among repeated elements in an individual body. Examples are the relation between forelimbs and hindlimbs, and among successive vertebra. Special homologies are body parts that correspond between species (*special* referring to species).

General homology is a subtler and more difficult relation, and it is seldom accurately reported in modern discussion (except Camardi 2001). General homology is reported in *Archetype* and also in *On the Nature of Limbs*, an 1849 lecture that was published as a small book and became extremely well known (Owen 1849). General homology is based on Owen's view that vertebrates are segmental organisms. A "vertebra" for Owen is not simply a bone but an entire bodily segment.

> I define a vertebra as *one of those segments of the endo-skeleton which constitute the axis of the body, and the protecting canals of the nervous and vascular trunks*: such a segment may also support *diverging appendages*. (Owen 1848: 81; emphasis in original)

Each of these segments is itself made up of parts that stand in definite relations to one another. Owen illustrates the relation between a real, natural "vertebral segment" and the ideal vertebra (an abstract segment made up of elements) in figures on facing pages (Owen 1849: 42, 43).

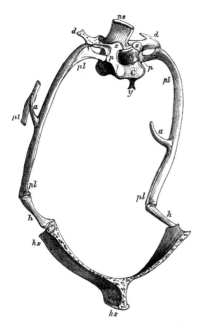

Natural skeleton-segment, 'osteocomma' or
'vertebra.' Thorax of Bird.

Fig. 2. A natural vertebra according to Owen (1849: 42). A vertebra is not a simple
bone but an entire segment, in this case a thoracic segment of a bird. Abbreviations refer
to the elements named in Figure 3.

The natural segment in Figure 2 is a bird thorax, including not only the
bone commonly named the "vertebra" but also the ribs and sternum. The
schematic representation of the generalized ("ideal") vertebral segment in
Figure 3 includes the neural arch above and the haemal arch below, and each of
the elements is named (neurapophysis, haemapophysis, centrum, etc.). Every
vertebrate body is made up of a series of these segments, and each segment
contains the same elements. The goal of the study of general homology is to
identify the bones of natural organisms both with respect to (a) which vertebral
segment the bone belongs to in the series of segments that makes up the body,
and (b) which *element of the segment* (neurapophysis, haemapophysis, etc.)
the bone represents in its respective segment. So we see that *Limbs* depicts not
one archetype, but two. Besides the renowned Vertebrate Archetype (Figure 4)
is the "ideal vertebral segment" (Figure 3). The former is a serial construction
of the latter.

Like almost all of his contemporaries, Owen accepted the vertebral theory
of the skull. The theory expresses the segmental nature of vertebrates: The

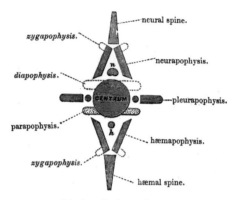

Ideal typical vertebra.

Fig. 3. The abstract archetype of the vertebral segment with its elements specified; Owen 1849: 43.

skull is composed of modified vertebral segments, just like the rest of the body.[27] *Limbs* explains how the ideal vertebral segment is modified differently in the skull and in the thorax (Owen 1849: 43). Cranial vertebrae have enlarged neural arches, to contain the brain, whereas thoracic vertebrae have enlarged haemal arches to contain the organs of circulation. Owen illustrates general homology by use of the vertebrate theory of the skull. He reports, "the basilar part or process of the occipital bone in human anatomy is the 'centrum' or body of a cranial vertebra" (Owen 1849: 4). The occipital bone is a part of the last vertebral segment of the skull, the so-called occipital vertebra. Owen goes on to describe how this implies that the item is therefore not really a "process" (an extension of a separate bone) but an independent vertebral element, *serially* corresponding with the centra of all other vertebra in that animal's body, and *specially* corresponding to independent bones in the bodies of cold-blooded vertebrates. It also has a special developmental relation to the notochord (*chorda dorsalis*) that identifies it as the centrum of that vertebral segment. Thus, the general homology of this bone extends in three dimensions: serially in the animal's body, specially in its relation to homologs in other vertebrates, and developmentally in the centrum's special relation with the notochord during embryogenesis. The complexity of this illustration shows the importance of general homology to Owen. It would be central to Owen's thoughts on species origins.

---

[27] Cruder versions had it that the skull was made up of the individual bones commonly called vertebrae. For Owen, "vertebrae" are entire body segments made up of elements, not simple bones.

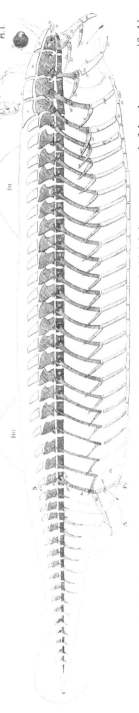

Fig. 4. The Vertebral Archetype. It is composed of a series of vertebral segments, each of which is in turn composed of the elements specified in Figure 3. The various stipplings indicate which element each bone represents. Owen 1849: Plate 1.

Richard Owen cautiously began to discuss naturalistic species origins in *Archetype*, and he expanded his comments in *Limbs*. His remarks were guarded and surrounded by pious rhetoric. This may have been why Darwin and some others didn't even recognize them. Unfortunately, Owen's conservative allies recognized them, and they attacked them harshly.

Owen's comments on species origins are closely intertwined with his concept of general homology and its role in the embryological formation of bodies. Development occurs under the influence of two general forces (or laws, or principles). One force is responsible for Unity of Type, and the other for diversity and adaptation. The structural force produces the "vegetative repetition" of identical elements in a body. It was originally described as a "polarizing force" such as that involved in magnetism or the growth of a crystal (Owen 1848: 171). In the vertebrate body it produces repeated vertebral segments along the front-to-back axis, serial homology. (Owen says that it is no surprise that a repetitious pattern of segments in vertebrates shows up best in their skeleton; the skeleton's mineral composition makes it closer to a crystalline structure.) The structural force also produces special homology, the identity of body parts between species (e.g., vertebrate limbs). The repeated elements produced by the structural force are modified by the adaptive force to serve diverse functions. These two forces act to some extent in opposition. The structural force dominates in lower life forms such as worms and starfishes. It also dominates in the lower forms within a class; the lowest vertebrates are closest to the Archetype.

Owen's two forces, structural and adaptive, account for both the diversity that we observe between species and the diversity that we observe in the body of a single organism.[28] Vertebrate bodies have diverse, specialized segments even though the segments are ideally identical. Both the adaptive and structural forces are at work *during the embryological development of the individual organism*. They operate "in the development of an animal body . . . during the building up of such bodies . . . in the arrangement of the parts of the developing frame" (Owen 1848: 171–172). The two forces are also responsible for the unity and diversity that we see between species. The homological identity of vertebrate limbs results from the similar action of the structural force acting during the embryological development of the dugong, and the mole, and the bat, and the horse, and the human. These limbs

---

[28] "Account for" in an abstract sense, of course, suitable to Owen's structuralism though not for Darwinian styles of evolutionary explanation.

are adaptively distinct because of the action of the adaptive force, also during embryological development.

Owen's concept of general homology led him to recognize a similarity between the adaptive specialization of segments in an individual body (a skull segment and a torso segment), and the adaptive specialization between different species of a common type (a bat's forelimb and a horse's forelimb). Just as a vertebrate body is a series of variations on the theme of the ideal vertebra, distinct species are variations on the Vertebrate Archetype. Owen believed that the recognition of these patterns, and the structural and adaptive forces that produced them, could lead to an understanding of the origins of new species on earth. This is not an "evolutionary theory" in the Darwinian or Lamarckian meaning of the term: Owen does not name a force that causes change. It can, however, be seen as a step toward such a theory.

Here is Owen's expression of that point in *Archetype*:

> To trace the mode and kind and extent of modification of the same elementary parts of the typical segment throughout a large natural series of highly organized animals, like the vertebrata: and to be thus led to appreciate how, without complete departure from the fundamental type, the species are adapted to their different offices in creation, brings us, as it were, into the secret counsels that have directed the organizing forces,* and is one of the legitimate courses of inquiry by which we may be permitted to gain an insight into the law which has governed the successive introduction of specific forms of living beings into this planet. (Owen 1848: 106; the asterisk refers to a footnote discussed in the paragraphs that follow)

This "law which has governed the successive introduction of specific forms" describes the naturalistic origin of species. (Species are interpreted as forms of course; Owen was a morphologist.) The paragraph describes how we might gain an insight into this law. First, we must examine how the adaptive force modifies each of the segments of an individual body during embryogenesis. Then we must notice how these within-body modifications vary between species, so as to adapt different species to different situations. The variation between species is produced by a difference in application of the same force that produces variation among the segments within a single body. *In order to learn how species (forms) originate, we must learn how bodies are built in the first place.* In other words, knowledge about embryological development is a step on the path to knowledge of the natural laws governing the origin of species.

Let us return to Owen's reference to the "secret counsels that have directed the organizing forces" (just the sort of expression that infuriated Huxley).

The footnote refers to an aphorism in the *Novum Organum* in which Bacon recommends the study of the growth of plants and the development of animals. Bacon says that it is strange that people study nature by examining her "finished products." It is as if one wanted to understand the work of an "artificer" but looked only at the raw material, and then the finished product. A wiser person "would rather wish to be present while the artificer was at his labors and carrying his work on" (Bacon 1960: 215). If we ignore Owen's baroque metaphors, we can see what he is getting at. Nature is "at her labors" in the production of organic form during embryological development, when the adaptive and structural forces are actually at work. If we can understand embryological development within an individual organism, we may come to understand how development *varies between species*. Then perhaps we can understand how development, at particular times in the prehistoric past, actually changed in such a way that new species (new forms) originated. To explain the origin of a new species is to explain how a new *form* came into being. The study of general homology is a step along that road.

This is an expression of what I call the *developmentalist doctrine* on the explanation of evolutionary change. The developmentalist doctrine conceives of evolutionary change between species as a change in the embryological processes that give rise to individual morphologies. To understand evolutionary change, one must first understand the processes of individual development, the ontogenetic processes that produce adult morphologies from single cells. These processes vary between species in a particular lawlike way, described by von Baer. They vary so as to produce adult morphologies whose taxonomic relations are expressed by the hierarchical Natural System. The evolution of one species into another species can only happen as a result of changes in the ontogenetic processes by which the species' morphology is constructed. If the ontogenetic processes don't change, the adult morphology cannot change. If our study of embryology and comparative morphology is successful, it may allow us to understand not only how existing morphologies *are generated* by related ontogenetic processes, but also how ontogenetic processes *can change*, resulting in new forms and new species. Only by understanding ontogeny is it possible to understand *changes in* ontogeny, and only by understanding changes in ontogeny can we understand the origins of new species.

In *Limbs*, Owen's discussion of naturalistic causes of species origins occurs prominently at the very end:

> To what natural laws or secondary causes the orderly succession and progression of such organic phænomena may have been committed we as yet are ignorant.

But if, without derogation of the Divine power, we may conceive the existence of such ministers, and personify them by the term "Nature," we learn from the past history of our globe that she has advanced with slow and stately steps, guided by the archetypal light, amidst the wreck of worlds, from the first embodiment of the Vertebrate idea under its old Ichthyic vestment, until it became arrayed in the glorious garb of the Human form. (Owen 1849: 89)

Owen is clearly not a creationist regarding species origins. He is pious, but he assumes the existence of secondary causes (natural laws) that brought new organic forms into being. He asserts that the study of homology will give us new insights into those laws.

The conservative backlash against Owen's work of the 1840s was focused both on the passages that discuss natural laws of species origin, and on his characterization of the two forces said to be responsible for the forms of bodies. His description of the forces changed in the three years between the original publications of *Archetype* and *Limbs*. Of special interest is Owen's reference to Platonic idealism. Neo-Darwinian commentaries often emphasize Owen's Platonism as an indication of the spiritualistic basis of his thought. The details of Owen's use of Platonism reduce its metaphysic significance.

In *Archetype*, the structural force is described in virtually materialist terms as an "all-pervading polarizing force" (Owen 1848: 171). The adaptive force corresponds to "the *ideas* of Plato . . . which [Plato] defined as a sort of models, or moulds in which matter is cast, and which regularly produce the same number and diversity of species" (Owen 1848: 172). Structure is materialist in origin, and adaptation is Platonic. But in *Limbs* the story suddenly changed. Owen identifies the *structural* force (not the adaptive force) as Platonic. The nature (or signification, or *Bedeutung*) of a limb is "that essential character of a part which belongs to it in its relation to a predetermined pattern, answering to the 'idea' of the Archetypal World in the Platonic cosmogony–" (Owen 1849: 2–3). In *Archetype* the adaptive force is Platonic, but in *Limbs* the structural force is Platonic. Owen gives no explanation for this flip-flop, but recent historical sleuthing has uncovered a probable cause. Owen reversed his Platonism in response to a challenge from the Oxbridge conservatives, who disapproved of Owen's willingness to explain Unity of Type by natural causes.

Owen's Platonic reversal followed a letter he received from the Cambridge conservative William Conybeare in 1848. Coneybeare suggested that Platonic idealism be relocated in Owen's theory, to replace materialist polarity as the structural force.

[Plato] meant the archetype forms of things, as they existed in the creative mind...To me the true...analogy seems to be the mind of a manufacturer about to produce his work; a shipwright his ship – an instrument maker his piano, or organ. (Coneybeare quoted in Rupke 1994: 202)

With the background of Coneybeare's letter, Nikolaas Rupke interpreted Owen's 1849 redeployment of Platonism as an attempt to "placate the powerful Oxbridge faction among Owen's supporters" (Rupke 1994: 204). Unity of Type had for years been a thorn in the side of adaptationist natural theology because of its apparent *lack of need* for theological underpinnings. Owen's polarizing force simply drove the thorn deeper. But if Owen could be convinced to relocate his Platonism to Unity of Type, transcendental anatomy might be brought under Plato's supernatural supervision. Owen cooperated.

This concession was not Owen's only response to Coneybeare's letter. *Limbs* did Platonize the Archetype, and Owen used the idealism to express his piety. However, Owen's real target of criticism in both books was the teleological biology practiced by Cuvier and the Bridgewater authors. He was championing structuralism; let Platonism fall where it may. Coneybeare had likened the Archetype to a plan in the mind of a designer, "a manufacturer about to produce his work; a shipwright his ship – an instrument maker his piano, or organ." The rigging of ships and the repeated patterns of musical keys are repetitious *for a purpose*. Was not nature the same? No. Owen's letter in reply to Coneybeare explains how Coneybeare is mistaken.

You will see, therefore, my dear Dean, that there are phenomena in animal structure, (and there are more in the Vegetable Kingdom,) that are not explained by the analogy of such seeming repetition-structures as you have adduced from works of human art. (Owen quoted in Sloan 2003: 60–61)

Unity of Type is *not* like an intelligent designer's plan, and merely attaching Plato's name to it doesn't change that fact. Owen did use Coneybeare's shipwright analogy in *Limbs*, but he stood it on its head. Owen's almost sarcastic use of the shipwright analogy is discussed in the following section.

Owen's discussions of the natural origins of species were guarded, but they were embedded in discussions of the Unity of Type, and the forces that controlled organic form both within and between vertebrate bodies. The mention of naturalistic species origins *in the context of* Unity of Type was a bold step. It was too bold for many conservative critics. The geologist and natural theologian Adam Sedgwick criticized Owen in print the following year, expressing grave doubts about Owen's pantheist tone. The *Manchester*

Guardian angrily condemned Owen in an editorial for his theologically unacceptable expression of "what is called THE THEORY OF DEVELOPMENT" (Richards 1987: 163 ff.). Unlike Coneybeare's letter, these were public condemnations. Stung, Owen ceased to publish on species origins. By the time he returned to the subject, Darwin's *Origin* had already scooped him.

4.5 ANTI-ADAPTATIONISM

On the Nature of Limbs is most frequently cited for two things: the Vertebrate Archetype and Owen's alleged Platonism. I consider the archetype to be very important, and Owen's Platonism to be a complete red herring.[29] A third feature of the book should not be overlooked. *Limbs* is intensely structuralist. It gives example after example of the failures of adaptationist, teleological reasoning. If one were to search for comparisons to *Limbs*, the *Bridgewater Treatises* is the last place to look. A far closer similarity is "The Spandrels of San Marco and the Panglossian Paradigm" (Gould and Lewontin 1979). Owen was certainly not as iconoclastic as Gould and Lewontin, but his defense of structuralism and critique of adaptationism was every bit as intense. The first refutation of adaptationism in *Limbs* includes Owen's reversal of Coneybeare's adaptationist shipwright analogy.

Owen first illustrates homology with detailed illustrations of the structures of vertebrate limbs: the mole, dugong, horse, bat, and human (Owen 1849: 4–9). Limbs are differently adapted, but they share common structure. His point is that the commonality of structure is *not* traceable to function – the exact opposite of the natural theologian's argument. Owen proves his point by using a piece of standard natural theological rhetoric, then standing it on its head. He compares limbs with human transportation inventions. His very first example is Coneybeare's shipwright. "To break his ocean bounds, the islander fabricates his craft, and glides over the water by means of the oar, the sail, or the paddle-wheel" (Owen 1849: 9). Owen proceeds to list a number of other human transportation devices. He then departs from the

---

[29] Rupke points out that Owen's archetype is unlike a Platonic Form (Rupke 1993: 243). Platonic Forms were taken to be the highest and most perfect exemplar, whereas Owen's Vertebrate Archetype is the most general and undifferentiated form. Given his flip-flop on its application, I see so little evidence of the metaphysical significance of Owen's Platonism that its deep interpretation is inconsequential. His idealism is no more metaphysical than Geoffroy's: a willingness to hypothesize beyond the limits of the conservative principle of the empirical accessibility of function.

natural theological game plan. Instead of pointing out similarities between natural bodies and human inventions, he indicates a crucial *difference* in the cases. Human ingenuity adapts an invention *directly* to its purpose, and does not make modifications to a common plan in order to produce a new invention. "There is no community of plan or structure between a boat and a balloon–" (Owen 1849: 10). Unity of type in the organic world does not correspond to a functional plan in the mind of a human inventor. The deepest truths in the organic world are those of Unity of Type and homology; these truths cannot be explained by teleological reasoning. Coneybeare's adaptationist shipwright is held up as the very model of adaptationist foolishness.

A second type of critique examines an adaptationist explanation of a morphological feature in humans, and it demonstrates that the feature occurs in other species by virtue of Unity of Type. However, the other species cannot possibly experience the benefit that has been attributed to the feature in humans. An example is the complicated pattern of unfused bones in the skull of an infant human. It had been proposed that the pattern of incompletely formed bones was an adaptation for the passage of the large human skull through the birth canal. Indeed the infant skull's compressibility does aid in birth.

> But when we find that the same ossific centres are established, and in similar order, in the skull of the embryo kangaroo, which is born when an inch in length, and in that of the callow bird that breaks the brittle egg, we feel the truth of Bacon's comparison of "final causes" to the Vestal Virgins, and perceive that they would be barren and unproductive of the fruits we are labouring to attain. (Owen 1849: 40)

The "principle of special adaptation" (i.e., adaptation in the individual species) fails to explain these structures. The centers of ossification are homologous. They are neither confined to one species nor confined to the species that have a functional need for them. Adaptationism fails again.

Owen's critiques of adaptationism are typically followed by pious passages in which Owen assures his reader that his skepticism about teleology does not imply irreligion. This was partly a sign of the times, but partly idiosyncratic to Owen. References to religion were common in popular scientific writings, even when the topic was not concerned with potentially controversial topics. Nevertheless, Owen is especially eager to demonstrate his piety in *Limbs*. He was aware that the Bridgewater conservatives had aligned structuralism with irreligion. Nevertheless, if the pious passages are read carefully, they are consistent with naturalistic species origins. Consider the pious conclusion to the skull discussion:

[I]f the principle of special adaptation fails to explain [the homologies], and we reject the idea that these correspondences are manifestations of some archetypal exemplar on which it has pleased the Creator to frame certain of his living creatures, there remains only the alternative that the organic atoms have concurred fortuitously to produce such harmony. (Owen 1849: 40)

The modern reader will see this as a false dichotomy. Surely special adaptation, divine archetypes, and sheer accident are not the only possible explanations. Ordinary natural laws are another! However, I propose that ordinary natural laws, secondary causes, are precisely how Owen believed that the "archetypal exemplar" became manifested in the world (see again Owen's dramatic concluding passage just quoted from *Limbs*). The divine creation of that exemplar at the beginning of time corresponded to the divine creation of the law of gravity. The First Cause created the law of gravity, and the organic laws by which individual species were caused to come into existence. Owen was pious in *Limbs*, even to the point of obfuscation – but he was no special creationist.

Owen was juggling different interests in these books, trying to express radical ideas without scaring the conservatives. However, if we look only at the pious packaging and not at the scientific content, we will lose sight of the actual theoretical contributions. Even Owen's description of the Platonic nature of the archetype in *Limbs* has a dual character, to please both theologian and scientist. His statement explaining the archetype begins "that essential character of a part which belongs to it in its relation to a predetermined pattern, answering to the 'idea' of the Archetypal World in the Platonic cosmogony." The sentence ends "–to which archetypal form we come, in the course of our comparison of those modifications, finally to reduce their subject" (Owen 1849: 2–3). Owen begins the sentence as a Platonist, and he ends it as an empiricist. It is easy to feel scorn for Owen's coddling of the natural theological sensibilities, but we should not allow it to distract us from his scientific intentions.

The naturalistic ideas about species origins that Owen expressed in the late 1840s do not look like "a theory of evolution" in the modern sense. His intention was not to propose a causal theory (a vera causa, an ultimate cause) at all, but merely to try to discern a natural law that described the coming into being of species. Species themselves were conceived morphologically, as forms. Owen's ambitions were to be like Kepler, not Newton.

Let us now examine how Darwin used the morphological facts provided by Owen and others in constructing his own theory.

4.6 DARWIN'S USE OF MORPHOLOGICAL TYPES

If one fact about Charles Darwin is beyond question, it is that he was an adaptationist. This can be seen in Darwin's earliest speculations on species change. In his 1837 Notebook B, he wrote that "the condition of every animal is partly due to direct adaptation & partly to hereditary taint" (Barrett et al. 1987: 182). Even at this early stage, Darwin saw the true nature of the species as embodied in its environmental fit; its hereditary structure was mere "taint." He remained an adaptationist even as the scientists around him in the 1840s were turning away from adaptation and toward structure. Owen's *Limbs* was a strong influence on this movement in Britain. Owen claims that homologs are "definable and recognizable under all their teleological modifications...through every adaptive mask" (Owen 1849: 41). For the structuralist Owen, adaptation was a "mask." For the adaptationist Darwin, hereditary structure was a "taint." Structure meets function.

Nevertheless, Darwin studied the structuralists and made good use of their work. Many commentators consider Darwin's theory to have been essentially complete in his *Essay* of 1844. Ospovat, in contrast, argues that the morphological and embryological research that Darwin read after 1844 was very important to the strength of his book, especially to his argument for the fact of common descent. Darwin's disinterest in embryology as an evolutionary *mechanism* may have made it possible for him to see more clearly the usefulness of structuralist biology as *evidence* for the sheer fact of common descent. He was rightly proud of his recruitment of this data. Embryology was his "pet bit" in the *Origin*, and the divergent von Baerian pattern of embryology "by far the strongest single class of facts in favour of change of forms." The "morphological or homological argument" was a close second (quoted in Ospovat 1981: 165). Darwin's own copy of *Limbs* has this marginal note: "I look at Owen's Archetypes as more than ideal, as a real representation as far as the most consummate skill and loftiest generalization can represent the parent form of the Vertebrata" (quoted in Ospovat 1981: 146). This reinterpretation of Owen's archetype as an ancestor was a very important step in Part 2 of the *Origin*, Darwin's argument for common descent.

The section on Morphology in Chapter 13 of the *Origin* acknowledges the work of Owen and other transcendental anatomists. Darwin begins with homologies of the vertebrate limb, and he points out that the same names

96

can be given to bones in widely different animals. (Darwin probably adopted this point from Owen's *Archetype*, its first and most dramatic statement.) His following two paragraphs allude to points from Owen's *Limbs*, although the spin is Darwin's own. First, on teleology:

> Nothing can be more hopeless than to attempt to explain this similarity of pattern in members of the same class, by utility or by the doctrine of final causes. The hopelessness of the attempt has been expressly admitted by Owen in his most interesting work on the "Nature of Limbs." On the ordinary view of the independent creation of each being, we can only say that so it is; – that it has so pleased the Creator to construct each animal and plant. (Darwin 1859: 435)

Darwin clearly credits Owen with proof of the failure of teleology in this passage, but he does so in a slightly backhanded way. Owen didn't just "admit" the failure of teleology in *Limbs*: he gleeful proved it! Darwin gains a definite advantage from Owen's point, as we can see by comparing the 1859 passage in the *Origin* to the *Essay* of 1844. In the earlier work, Darwin had recognized that Unity of Type *can* be explained by common descent, but he had not recognized (at least not openly) that Unity of Type *cannot* be explained adaptively. As an adaptationist himself, he may not have been on the lookout for the breakdowns of adaptationism. Owen was on that lookout. His success shows Darwin how to refute one particular brand of adaptationism, the brand that special creationists had relied upon. Owen (and Darwin following him) refutes *special* adaptationism, the kind that is applied directly to species without recognition of the constraints of common descent.[30]

The next paragraph contains Darwin's flourish of transforming the archetype to an ancestor:

> If we suppose that the ancient progenitor, the archetype as it may be called, of all mammals, had its limbs constructed on the existing general pattern, for whatever purpose they served, we can at once perceive the plain signification of the homologous construction of the limbs throughout the whole class. (Darwin 1859: 435; note that Darwin speaks of the ancestral mammal rather than the ancestral vertebrate.)

---

[30] It might be thought that Darwin was a special adaptationist also, in that natural selection works within species (or population) and not at higher taxonomic levels. Although this is true, traits produced by natural selection are not restricted to the species level for Darwin. They are generally passed on to daughter species, and they can even become a part of the "type" (Darwin 1859: 206). So the existence of an adaptive trait does not demonstrate that the trait was created in and for the species that holds it. This is why special adaptationism is an argument for special creationism.

Owen was so strongly associated with the concept of the archetype by this time that there was no need to cite him. Darwin did consider Owen to be a species fixist when he wrote the first edition of the *Origin*, so his use of the expression "plain signification" may have a jibe. *Signification* had been the term Owen used at the beginning of *Limbs* to translate the German word *Bedeutung*, a term which he had borrowed from Lorenz Oken's first publication of the vertebrate theory of the skull (Oken 1807). Owen thought that the *Bedeutung* of limbs was their position in terms of general homology; Darwin believed it simply to be the ancestor's limb.

Darwin next discusses serial homologies, including the vertebral theory of the skull. On this topic he again deploys one of Owen's critiques of teleology, directly citing Owen's refutation of the teleological explanation of infant skull structure:

> Why should the brain be enclosed in a box composed of such numerous and such extraordinarily shaped pieces of bone? As Owen has remarked, the benefit derived from the yielding of the separate pieces in the act of parturition of mammals, will by no means explain the same construction in the skulls of birds. (Darwin 1859: 437)

Then there is the Darwinian version of Owen's interaction of the structural (vegetative) force with the adaptive force. Darwin subsumed the structural force under heredity and the adaptive force under natural selection:

> An indefinite repetition of the same part or organ is the common characteristic (as Owen has observed) of all low or little-modified forms; therefore we may readily believe that the unknown progenitor of the vertebrata possessed many vertebrae; ... consequently it is quite probable that natural selection, during a long-continued course of modification, should have seized on a certain number of the primordially similar elements, many times repeated, and have adapted them to the most diverse purposes. (Darwin 1859: 437–438)

Owen's principle of vegetative repetition followed by adaptive specialization has always been an important part of morphology. It has taken on a new importance in evo–devo, with crucial aspects of evolution attributed to duplication and subsequent specialization of entire gene families (Gerhardt and Kirschner 1997).

Many modern commentators misperceive Darwin's reliance on the morphological (and indeed the rhetorical!) work of Owen. They do so for an interesting reason. Synthesis Historiography (SH) allows no room for a structuralist contribution to evolutionary science. Darwin's genuine respect for Owen's results is therefore misread as disapproval.

Mary Winsor discusses how difficult it is for modern thinkers to understand pre-Darwinian naturalist thought.

> [M]any eminent and imaginative men, while disbelieving in a genetic connec-
> tion between species, did use classification to express their belief in natural
> relationships. At first we might expect that those scientists must have had some
> explanatory system which could make classification meaningful, as evolution
> makes it meaningful today. (Winsor 1979: 4).

They did not have such an explanatory system, and so it is hard for us to appreciate what they thought they were up to. The same problem applies to pre-Darwinians such as Owen, who *did* believe in broadly genetic re-lations among species but could not conceive of natural laws that would explain species origins. This interpretive problem is partly caused by SH it-self, and the conceptual apparatus that was introduced to help us understand neo-Darwinism. An example is Ernst Mayr's proximate–ultimate distinction, introduced at the very beginning of SH (Mayr 1961). Proximate causation involves the processes in an individual's lifetime, and ultimate causation in-volves the historical origins of the characters of an individual organism. How do we characterize the Natural System, as seen by (say) Strickland in the 1840s? Or the Vertebrate Archetype, as seen by Owen in 1849? Neither prox-imate nor ultimate causation captures the meaning of these concepts. They do not directly involve *causation* at all.

Proximate–ultimate is not the only way to categorize scientific beliefs. A different dichotomy is between phenomenal laws and causal laws. Phenome-nal laws, sometimes called geometric laws, describe observable patterns but do not attribute causes. Causal laws are aimed at vera causa, the true underly-ing causes that explain phenomenal laws. Kepler is the model of the scientist in search of geometric laws, and Newton the model of the genius who can achieve a causal law (Ruse 1979; Hull 1983). If we view the realistically interpreted Natural System and Owen's Vertebrate Archetype as geometric laws of the organic world, pre-Darwinian thinkers start to make a bit more sense to us.

Practitioners of SH do not view the pre-Darwinian typology as a matter of geometric laws, however. They apply the proximate–ultimate distinction instead; theories are either proximate or ultimate. Evolutionism and special creationism are both ultimate theories. What is morphological typology? It is clearly not proximate, so it must be ultimate. It is clearly not evolutionary, so . . . it must be creationist. If it is not creationist, it must be vacuous. This dilemma – either creationist or vacuous – appears in an early analysis by David Hull:

> As long as one believed in God, and these plans could be interpreted literally as thoughts in the mind of the creator, then such explanations had some explanatory force, but if reference to God is left out of the explanatory picture, then all that is left are the plans. Rather than being explanations, the existence of such "plans" calls for explanation. (Hull 1973: 74)

The typological theorists could not be interpreted as contributing to the data from which Darwin inferred evolution, because they were proposing antievolutionary (creationist) ultimate–causal theories. Or, if they were not proposing causal theories, then their writings were vacuous.

Several neo-Darwinian critiques of typology have called attention to the passage on teleology from the Morphology section of the *Origin* quoted here in Section 4.6, in which Darwin discussed the failures of teleology. Darwin treats Owen with respect in the passage and credits him with refuting teleological finalism. Authors find it difficult to accept Darwin's acknowledgment of Owen, but in order to play down Owen's contributions, they are forced either to misinterpret Darwin or to disagree with him. I will reproduce the passage here, along with a clause that Darwin added in 1866.

> Nothing can be more hopeless than to attempt to explain this similarity of pattern in members of the same class, by utility or by the doctrine of final causes. The hopelessness of the attempt has been expressly admitted by Owen in his most interesting work on the "Nature of Limbs." On the ordinary view of the independent creation of each being, we can only say that so it is; – that it has pleased the Creator to construct all the animals and plants in each great class on a uniform plan [; but this is not a scientific explanation]. (Darwin 1859: 435; bracketed passage added to the 1866 fourth edition)

Here is Ernst Mayr's interpretation of the passage:

> The idealistic morphologists were completely at a loss to explain the unity of plan and, more particularly, why structures rigidly retained their pattern of connections no matter how the structures were modified by functional needs. As Darwin rightly said "Nothing can be more hopeless than to attempt to explain the similarity of pattern in members of the same class, by utility or by the doctrine of final causes." (Mayr 1982: 464)

The first sentence of Mayr's passage reiterates Hull's claim of the vacuity of idealism. The second sentence, amazingly, claims that Darwin was refuting the idealist morphologists, but it quotes a passage that Darwin had in fact *credited to* an idealist morphologist (Owen) as a refutation of Bridgewater adaptationism! Mayr is here claiming for Darwin a discovery that Darwin *himself* credits to Owen (and does so in the immediately following sentence).

Peter Bowler is more accurate than Mayr in discussing this passage. He acknowledges that Darwin credits Owen with refuting teleology, but he thinks Darwin is too generous. He finds it "curious" that Darwin credits Owen "without mentioning that [Darwin's] theory made equal nonsense out of Owen's own explanation" (Bowler 1977: 37). The "nonsense" refers to Owen's alleged Platonism. Bowler apparently considers Owen's Platonism as an ultimate–causal explanation, in direct competition with an evolutionary explanation. It is quite clear from the *Origin* that Darwin did not interpret Owen in this way.

The simple evolution-or-creation motif of neo-Darwinian commentary was greatly enriched in 1983 by a paper in which Hull recognized idealism as a distinct approach to biology and not merely a subtype of creationist natural theology. Evolutionism, creationism, and idealism were now seen by Hull as the contending doctrines, with "reverent silence" a fourth option (Hull 1983: 63). Idealists are here described in two ways. Only some of them were essentialist antitransmutationists (e.g., Agassiz and Dana). Nevertheless, even though Hull no longer equates idealism with creationism, he still reads Darwin as disapproving of Owen's idealism, and of doing so in the *Origin*. He reports that Darwin had described Owen's *Nature of Limbs* only as "interesting" in the first edition. In later editions "Darwin was more candid, concluding his discussion of Owen with 'but this is not a scientific theory'" (Hull 1983: 71).

The passage to which Darwin appended "not a scientific explanation" speaks for itself. The insertion into the fourth edition is placed in brackets in the quotation given here in an earlier paragraph. The claim of "not a scientific explanation" is clearly not attributed to Owen's *Nature of Limbs* but rather to "the ordinary view of the independent creation of each being," which is to say, to special creationism. Darwin does not attribute creationism to Owen in this passage, even in the first edition. By the time he inserted the "not an explanation" clause, he had already admitted that Owen was *not* a special creationist. Darwin did not claim that Owen's explanation was nonscientific.

The view of typology as *either* vacuous *or* creationist was common thread in neo-Darwinian historical commentary. As these examples show, it was not Darwin's opinion. Darwin recognized the importance of Owen's work even though he had (incorrectly) considered Owen to be a species fixist. Owen's striking illustrations of the failure of teleology were weapons in Darwin's hands against the special adaptationists of the Bridgewater generation. Owen's archetype was ready-made for Darwin's conversion to an ancestor. Huxley and Darwin commiserated with one other about the obnoxiousness of Owen's rhetoric, but that rhetoric was not even mentioned in the *Origin*. Modern

of a species can be inherited by its descendant species does not prove that adaptations are "higher" than structure. To a structuralist, every adaptive modification is a modification of a structure. The ancestral bat may have passed on its adapted wing as part of the type of later bats, but that wing was a vertebrate forelimb *before* it became a wing. Its homological identity (its forelimbness) was passed on to descendents just as much as its adaptive modification (its wingness). Moreover, its homological, structural identity *came first!* Darwin could argue that the earlier structure had itself been adaptively shaped in an even earlier ancestor, but there is no end to this argument. Even if every currently existing organ was shaped by adaptive change, that adaptive change happened *to a structure* that preexisted the adaptive change. *Which came first: form or function?* This truly is a chicken-or-egg question.

For Darwin, Unity of Type is a mere by-product of adaptive modification. Natural selection produced that modification. Unity of Type is not causally involved in the process of change. It receives a sort of explanation by exclusion: Unity of Type is the sum total of ancestral characters that *were not* modified by natural selection. I call this the *residual concept* of homology and Unity of Type: It will recur in modern neo-Darwinian thought.

4.9 A STRUCTURALIST EVOLUTIONARY THEORY?

But what alternative could there be? How could Unity of Type be causally involved in a theory of evolutionary change? Here's how.

A structuralist evolutionary theory involves a distinct causal process and a distinct explanatory goal. The explanatory goal is the traditional goal of morphology: the explanation of form. The causal process is harder to explicate. Keep in mind that structuralists had recognized the geometric parallels among various aspects of morphology. One aspect was the comparative morphology of adults, with its pattern of groups within groups. Another was the branching pattern of von Baerian embryology. A third pattern (less robust) was the origin of forms in the fossil record. Notice that one of these patterns is more directly causal than the others. It is embryology: The heterogeneous form of an adult gradually emerges by differentiation from an earlier, more homogeneous embryo. What would be required for a *new form* (a new species) to appear on the earth? The answer is this: "In order to achieve a modification in adult form [something] must modify the embryological processes responsible for that form" (Horder 1989: 340).

For structuralists, species origins were the *origins of new forms.* Form arises in ontogeny by means of the form-generating process of embryogenesis.

Therefore, the origins of new forms must come about through *changes in the process of embryogenesis.* Owen had sketched some ideas about how form generation occurred in the embryo through the interaction of the adaptive and structural forces. This process occurred in a different way in different species, but apparently it occurred similarly in species whose ontogenies were most similar. If we could understand embryogenesis better, we might then be able to understand how its elements could be modified. Owen was never able to fill in the blanks and discover either (a) the mechanics of form generation in the embryo, or (b) how those mechanics could be modified.

One difficulty in understanding early structuralist thought about species origins is that modern evolutionary thought is centered on what we call a *mechanism* of evolution: natural selection. There was nothing analogous in structuralist thought. The closest thing to a mechanism was precisely this: the means by which the process of embryogenesis can be modified. The goal was the explanation of the varieties of form throughout evolutionary time *in terms of* the ontogenetic generation of form, together with an understanding of how the processess of ontogeny can vary. This is the developmentalist doctrine. Understanding species origins would require some understanding of embryogenesis. Owen's discussion of the adaptive and structural forces was a step in that direction, but there was obviously a long way to go.

Darwin's theory did not provide an answer to this question. He did not explain how the embryogenesis of one species can be modified to produce a different form. However, he argued very cogently that evolution had actually occurred, and he proposed a mechanism (natural selection) that was supposed to explain evolutionary change *without* explaining how embryogenesis could be modified. Instead of untying the Gordian knot, Darwin sliced through it.

4.10 HOW DARWIN DIFFERED

We haven't yet discussed Darwin's explanatory goal in a way that contrasts with the structuralists' goal of explaining form. It might be argued that Darwin did explain form; he just did so by use of natural selection. However, this is not the kind of explanation of form qua form that the morphologists sought. Darwin's explanatory goal was quite different – but I believe that appreciating its difference will allow us to recognize Darwin's accomplishment in a new light.

I propose that Darwin's explanatory goal was not the explanation of form; it was the explanation of change. He did not have the morphologists' ambition to unify embryology, taxonomy, and comparative anatomy, or to trace form back to its earliest ontogenetic beginnings in homogeneity. Natural selection operates on adult characters *irrespective* of their ontogenetic origins. Darwin was not interested in the *origin* of form. He was interested, instead, in how form changes. He was equally interested in how *any* characteristic of a species changes. Organic form was nothing special; it was just one characteristic among others. Compare Darwin's discussion of the archetype-ancestor with Owen's original discussion of the Vertebrate Archetype. Darwin spoke of the ancestor not of vertebrates, but of *mammals*. The ancestor-archetype of mammals already had a limb! Owen's Vertebrate Archetype did not. Owen had identified the limb with nonlimb elements of vertebral segments. Darwin identified the limb with previous limbs. When Darwin transformed the archetype into an ancestor, the archetype concept was disempowered; it lost its morphological explanatory force. Darwin didn't care about that loss, because he was not in the morphologists' business of explaining form. For Darwin, the ancestor was merely a hypothetical starting point from which change occurred. To a morphologist, the archetype is a theoretical construction that plays a role in the explanation of form. A morphologist would not trade in an archetype for a mere ancestor; to do so would be to abandon the goal of morphological explanation.

Darwin's approach to evolution allowed him to sidestep the problem of form entirely. This new approach has definite advantages. For example, it operates equally well on characters for which the ontogenetic origins are unknown! This offers a tremendous broadening of the scope of evolutionary biology. Emotions and instinct can be studied, even though we have no idea how they arise by ontogeny in the individual!

Darwin's new explanatory goal was change qua change. Each Darwinian explanation begins by *assuming* the existence of an ancestral population; the characteristics of that population are not themselves explained. Darwin does not feel responsible to explain the ontogeny of the characters within the ancestral population: He is not a morphologist. The rich texture of morphological, embryological, and taxonomic unity is an epiphenomenon, merely a side effect of the operation of the Darwinian adaptive engine. Many of Darwin's examples illustrate this technique, as for example his discussion of the instinct of "slave-making" in ants (Darwin 1859: 223–224). He describes the differences among several species of ants, including some that carry off pupae of others species, some that rely partially on "slaves" of other species, and some that rely entirely on the "slaves." This alone established the plausibility of a selective explanation of the instinct. Darwin feels no need to explain the acquisition of the habit (or instinct) in individual ants. The ontogeny of the instinct is irrelevant to its phylogeny, as far as Darwin is concerned. Darwinian explanations accounted for the evolution of a trait without even considering its development within an individual species member.

This is a difficult result for a morphologist to swallow. Owen never swallowed it – nor did Huxley. Huxley in 1876 described Darwin's approach to cosmology as one that "assumes that the present state of things...has been evolved by a natural process from an antecedent stage, and that from another, and so on; and, on this hypothesis, the attempt to assign any limit to the series of past changes is, usually, given up" (Huxley 1893b: 50). Huxley didn't reject Darwin's goals, but he still retained the evolutionary morphologist's desire to explain the forms of organisms, and to do so by means of an understanding of how form was generated in the individual. "[E]volution is not a speculation but a fact; and it takes place by epigenesis" (Huxley 1893a: 202). Even after Darwin, morphologists often indicate that they are still interested in the old tradition of explaining form, and that natural selection, with its ability to explain changes of form, is a poor substitute. Embryologist E. E. Just complained that genetics and selection could explain why populations of flies had more or fewer bristles on their backs, but it couldn't explain how a fly constructed its back in the first place (Harrison 1937: 372; Gilbert et al. 1996: 361).

If I am correct in this analysis, Darwin invented not only a new causal mechanism but also a new explanatory goal for it to accomplish. The goal seems almost self-evident today; the structuralist goal is more in need of explanation. However, the modern assumptions of neo-Darwinism (e.g., that ontogeny is irrelevant to phylogeny, or that typological thinking is unscientific) make it difficult for us to recognize the theories that were alternatives to Darwin's in his own day. They were not antiscientific or mystical. They simply had different explanatory goals from Darwin's. Those goals are renewed in evo–devo.[31]

---

[31] Sections 4.8 and 4.9 in Chapter 4 were more strongly biased by my interests in modern evolution theory than other portions of the chapter. The quotation from Horder was not intended by him to apply to an evolutionary theory of the mid-nineteenth century. Tim Horder's 1989 paper argued for the integration of development into modern evolutionary biology. I replaced his statement that "evolution must modify the embryological processes" with "[something] must modify the embryological processes" in order to express what I believe was on morphologists' minds in the 1850s. The comment attributed to E. E. Just was from the 1930s, in the context of an embryologist's criticism of Mendelian genetics. I thought it illustrated well how morphologists might have reacted to Darwin's replacement of the explanation of form with an explanation of change of form by means of natural selection. I jiggered these reports in order to show how persistent the contrasts are between adaptationist and structuralist approaches to evolution, contrasts that I see as continuing through time and through very great changes in our knowledge about the facts of biology.

5

Evolutionary Morphology: The First
Generation of Evolutionists

5.1   THE PROGRAM OF EVOLUTIONARY MORPHOLOGY

The research program of evolutionary morphology comprised "the first generation of evolutionary biologists" (Bowler 1996: 14). It arose soon after the
1859 publication of the *Origin*. By the turn of the century, it had almost
completely died out. Most of its adherents had turned away from phylogenetic studies and toward the experimental study of embryology and genetics.
Depending on one's theoretical commitments, evolutionary morphology was
either the last gasp of the metaphysically flawed program of idealist morphology, or else the birth of the promising program of evolutionary developmental
biology. I favor the latter interpretation. Peter Bowler favors the former. Both
points of view are discussed in this chapter.

Peter Bowler has written three extraordinarily valuable books on the history of this period (Bowler 1983, 1988, 1996). He was among the first historians to examine the contrast between modern neo-Darwinian theory and
the views on evolution that preceded the Evolutionary Synthesis. The late
nineteenth century has received much less historical attention than the period
just prior to the *Origin*. Bowler's work fills an important historical gap, and
it presents us with an intriguing and (at first sight) surprising report on how
very different the evolutionary theorizing of that period was from that of the
mid-twentieth century. I am extremely indebted to this work, but I have one
important disagreement with its author. I consider Bowler's histories of the
period to be more important than he considers them. Bowler takes himself
to be examining, in large measure, the methodologically and metaphysically
flawed theories that preceded our modern scientific understanding. I consider those same theories to be legitimate and productive attempts to solve
problems that have been marginalized during most of the twentieth century,
problems that are once again receiving scientific attention. The contrast in

historical perspective can be seen between Bowler's *Life's Splendid Drama* and Brian Hall's *Evolutionary Developmental Biology* (Bowler 1996; Hall 1999a). Hall is a prominent and historically sophisticated practitioner of evo–devo. He treats the evolutionary morphologists as his intellectual ancestors. In contrast, Bowler takes an almost apologetic stance toward his study of the period. He considers the evolutionary morphologists to be irrelevant to modern science and of interest only to historians:

> If we wish to tell our story as a triumphant advance toward modern evolutionism, we may be justified in ignoring the role of morphology in the Darwinian revolution. But if we want to understand what evolutionism actually meant to late nineteenth-century biologists... we ignore it at our peril. (Bowler 1996: 14)

Bowler considers the evolutionary morphologists to be relics of the past with no surviving descendents; Hall (a currently active researcher) considers himself a descendent of the evolutionary morphologists.

Why this contrast? It follows from the differing views of Bowler and Hall on the nature of contemporary evolutionary biology. Bowler represents a neo-Darwinian perspective, and Hall an evo–devo perspective. I argue that the two views differ so markedly because neo-Darwinian evolutionary theory does not recognize the *explanation of form* as an explanatory goal. From the neo-Darwinian perspective, evolutionary morphology was a dead end, and Bowler reports this fact. However, in a historical twist, the emergence of evo–devo in the 1990s embodies a return to the goal of explaining form. This chapter concludes with a report on how recent work in evo–devo represents a rebirth of the goals of nineteenth-century evolutionary morphology. Even though Bowler doesn't recognize the fact, his studies of evolutionary morphology reveal the intellectual ancestors of cutting-edge evolutionary science.

5.2 EVOLUTIONARY MORPHOLOGY AS NON-DARWINIAN AND AS DARWINIAN

The label "Darwinian" today implies a much more specific set of evolutionary commitments than it did in the 1860s. Virtually all of the evolutionary morphologists considered themselves Darwinians. Bowler has identified nineteenth-century evolutionism as *The Non-Darwinian Revolution* because of the nature of evolutionary morphology. I agree that evolutionary morphology was not *what we now call* Darwinian. Our modern meanings have been shaped and sharpened by the Evolutionary Synthesis. Today a Darwinian approach to evolution must emphasize adaptation and natural selection, not

Unity of Type. But let us examine the tradition in its own terms, which include its commitment to what was then considered Darwinism.

Lynn Nyhart groups the German morphologists of the nineteenth century into six generational cohorts (Nyhart 1995: 20 ff.). The first cohort included the Naturphilosophen and their contemporaries. The second through the fourth cohorts were professionally active when Darwin's *Origin* appeared. The fifth cohort received its education in the tradition of evolutionary morphology, but members also began programs in experimental embryology. The sixth and last cohort, at about the turn of the twentieth century, rejected evolutionary morphology and turned toward experimental studies of embryology and genetics.

Each cohort had its methodological idiosyncrasies. The Naturphilosophen were elaborate if not florid in their discussions of methodology and idealist metaphysics. This style fast fell from favor. The second cohort, in the 1820s and 1830s, began to use *Naturphilosophie* as "a convenient label of derogation" (Nyhart 1995: 44). Open discussion of metaphysics faded into the background as German scientists began to adopt the same cautious inductivism as their British colleagues. Metaphysics again emerged in scientific discussion with the fourth cohort, after evolutionism had been accepted. Now, however, the commitments were to the "antimetaphysics" of materialist reductionism. The cautious inductivism of the early part of the century had smoothed the transition to the radical new theory of evolution and its materialist metaphysics. The research goals of morphology persisted through the changes in metaphysical commitment. Philosophical doctrines began with metaphysical idealism, which gave way to inductivist silence about metaphysics, which in turn gave way to metaphysical materialism. The morphological explanatory goal remained. The goal was the explanation of organic form, and the discovery of the relation between ontogeny and the Natural System.

The three cohorts that dealt with Darwin's challenge all believed in a lawlike connection between ontogeny and the Natural System (Nyhart 1995: 139). The cohort that first rejected Naturphilosophie (the second cohort overall) included von Baer and Johannes Müller. To them, the relation between ontogeny and the Natural System was conceived as an underlying force "that had the unique property of moving the organic world to ever higher levels of complexity and perfection" (Nyhart 1995: 140). This force was generally but vaguely associated with an intelligent creator. The third cohort, including Kölliker and Leuckart, was more fully imbued with inductivist caution. They avoided discussion of metaphysics, and they avoided any abstract principles except those that could be justified as heuristically useful. The fourth cohort included Haeckel, Gegenbaur, and other early evolutionary morphologists.[32] Nonmaterial forces had vanished. Adaptation and heredity were seen as the two important causes of evolution. Each cause was empirically demonstrable, and each was seen to be reducible to the deterministic material laws of chemistry and physics. Idealism *as a metaphysical doctrine* was denounced. Nevertheless, they retained the goal of the explanation of form, and the typological explanations that had earlier been tied to idealist metaphysics. So the explanatory use of ideal types was retained in the absence of metaphysical idealism.

Evolutionary morphologists recognized Darwin as the innovator of the concept of community of descent, and of natural selection as the explanation of adaptation. These two Darwinian concepts offered new ways of conceiving the old dichotomy of Conditions of Existence and Unity of Type. Natural selection provided a much-appreciated materialist account of adaptation. It allowed the rejection of old-fashioned teleology without rejecting adaptation itself (see my discussion of Leuckart in Chapter 3, Section 3.4.3). The reality of natural selection did not imply that adaptation must dominate the study of evolution – far from it. Common ancestry contributed to the explanation of Unity of Type, and evolutionary morphology was a structuralist program. Even though a twentieth-century Darwinian must be committed to the centrality of adaptation, in the nineteenth century, structuralists such as Haeckel and Huxley loudly proclaimed their Darwinism.

Mario Di Gregorio explores the transition from nonevolutionary to evolutionary morphology, and how the evolutionary morphologists saw themselves as Darwinian. "The Haeckel–Gegenbaur image of evolution derives ultimately from idealistic morphology, connecting both with German *Naturphilosophie* (more visibly in Haeckel) and with an Owen-echoing typology (more visible in Gegenbaur), transmuted into a new evolutionary morphology" (Di Gregorio 1995: 248). These connections look highly suspicious from the standpoint of Synthesis Historiography. However, the evolutionary morphologists carefully distinguished between the metaphysically pernicious commitments of their predecessors, such as metaphysical idealism and teleology, and those that could be understood on a materialist basis, such as recapitulation and other structuralist patterns of relatedness. "The 'embarrassment' was . . . not so much typology as [metaphysically] idealistic typology . . . the new theoretical instrument was supplied by Darwin's concept of community of descent, which allowed old-style . . . typology to be recast as evolutionary typology" (Di Gregorio 1995: 253).

---

[32] Gegenbaur actually fit somewhere between the third and fourth cohorts.

Di Gregorio's term *evolutionary typology* clashes with modern vocabulary. *Darwinian* typology is a definite contradiction in terms. On Mayr's account, Darwin had replaced typological thinking with population thinking. How could typologists possibly consider themselves to be Darwinians? Coleman quotes the 1870 statement in which Gegenbaur first explains his Darwinian typology.

[Darwin's] theory allowed what previously had been designated as *Bauplan* or *Typus* to appear as the sum of the structural elements of animal organisation which are propagated by means of inheritance, while modifications of the structures are explained as being adaptations. Inheritance and adaptation are thus the two important fulcra which render intelligible both the multiplicity and the unity of organisation. From the standpoint of descent theory, the relatedness of organisms loses it figurative meaning. Whenever we encounter through the use of precise comparison demonstrable agreement in structural organisation, this indicates common ancestry founded on inheritance. (Gegenbaur 1870, translated in Coleman 1976: 162).

Di Gregorio offers an interesting discussion of how morphologists could read the *Origin* as a legitimation of evolutionary typology. Darwin's application of natural selection to embryology absolved the "inactive" embryo from selective forces, and thus it allowed heredity to control the earliest stages of the developing type. Di Gregorio describes the Darwinian typologist as taking "natural selection as a causal mechanism that accounts for the *reality* of type-phenomena (rather than, for example, as a theory that makes the reference to types redundant)" (Di Gregorio 1995: 260). The evolutionary morphologists considered both heredity and natural selection to be evolutionary mechanisms, but they had little interest in the causes of adaptation. Selective forces were merely the background conditions that produced variation in inherited form. Natural selection was considered to be an external force, and the morphologists were more interested in the organisms' own internal structure. "Since the forces determining the changes of the organism lie outside [the organism] or for the most part are to be sought there, their [consideration] lies beyond our responsibility" (Gegenbaur 1870, translated in Coleman 1976: 172).

As the passage shows, the modification of *inherited structure* is Gegenbaur's primary interest. Evolutionary modifications to conserved structure are caused by natural selection, but the adaptive significance of these changes is not a part of the morphological program. The type or bauplan is not refuted but reinterpreted in this version of Darwinism. Gegenbaur shows no yearning for a dualist metaphysics in which types subsist in a world of ideas, or in the mind of God. Gegenbaur's disinterest in adaptation did not stop him from considering himself a Darwinian. Once we recognize the differing significance of the term *Darwinian* then and now, there is little reason to quibble over its use.

5.3 THE BIOGENETIC LAW

Darwin recognized the importance of embryology to classification, and he believed that his theory could account for this importance. Because embryos were usually protected from the environment, they were less affected by natural selection, and so they were relatively less modified from their ancestral form. This can be seen in his *Sketch* of 1842: "The natural system being on theory genealogical, we can at once see, why foetus, retaining traces of the ancestral form, is of the highest value in classification" (Darwin 1909: 45). This fact, suitably elaborated, became the central and most notorious aspect of the program of evolutionary morphology. Haeckel apparently coined the terms *ontogeny* and *phylogeny* in order to be able to express the law succinctly: *ontogeny recapitulates phylogeny*. According to Gould, the application of re-capitulation to evolution was rediscovered independently at least four times in the decade after the *Origin*, by Fritz Müller, Haeckel, and the paleontologists Edward Drinker Cope and Alpheus Hyatt (Gould 1977: 70). Recapitulation made embryology a lens through which one could inspect phylogenetic history.

The crucial issue for evolutionary recapitulation is that the early stages of embryos are taken to resemble the *adult* stages of ancestral organisms. This will happen only when evolutionary changes occur late in ontogeny, especially by "terminal addition," the addition of new stages onto a mature adult. Darwin believed that this was generally true. His reasoning was that evolutionary changes are adaptively driven (either by use-inheritance or natural selection), and the environment puts more adaptive demands on adults than on early embryos. If evolutionary changes did occur in early ontogenetic stages, the embryo would not resemble ancestral adult forms. The biogenetic law could still be applied if one could tell the difference between the embryonic traits that had been terminally added and those that had been interpolated early in ontogeny. The fortunes of recapitulationist theory, and of evolutionary morphology in general, were to hinge on the ability of morphologists to tell the difference. Haeckel called the first "palingenic," and the second "cenogenic" traits. Recapitulationists from the very start had recognized this problem. If the morphologist misinterpreted a cenogenic trait as palingenic, absurdity would result. Birds could not have evolved from an ancestor that lived, as an adult, encased in an egg.

The first published description of recapitulation was by Müller in 1861, in a book translated into English under Darwin's sponsorship. Müller distinguished two modes of evolutionary change. The first was when a descendent form was modified *"by deviating sooner or later whilst still on the way towards the form of their [adult] parents."* This would give rise to cenogenic change, in which embryological forms would mislead the recapitulationist. The second mode was as follows:

> *by passing along this course without deviation, but then, instead of standing still, advanc[ing] still further. . . . In the second case the entire development of the progenitors is also passed through by the descendants, and, therefore, so far as the production of a species depends on this second mode of progress, the historical development of the species will be mirrored in its developmental history.* (Müller 1869: 111–112; emphasis in the original)

Müller also believed that the embryological evidences of ancestry would eventually become "effaced" as ontogeny would gradually tend to take a "straighter course" from the egg to the adult (a process later called *condensation*) and that they would be disguised by the adaptations that arose in free-living larvae (Müller 1869: 114). Even with all of these confounding variables, embryonic forms have the potential to reveal ancestral forms. Müller illustrated his notions by showing how they inspired his discovery of the nauplius larvae of shrimp. He constructed a phylogeny in which the nauplius form represented the ancestor of all crustacea. Müller's discovery of the nauplius larvae shows that the recapitulationist thinking was not oriented only at ancestral reconstruction; it had implications for contemporary research.

Recapitulation is seldom treated with sympathy in modern times. Modern thinkers see von Baer's laws as sufficient grounds to oppose the recapitulationist identification of embryonic stages as ancestral adults. If von Baer's laws refuted the Meckel–Serres law of a linear succession of adults, wouldn't that apply as well to recapitulation? No. As elegant as von Baer's laws are, they are false of the embryology of complex organisms. Ernst Mayr explains why by quoting evolutionary morphologist Frank (Francis Maitland) Balfour. Balfour was a prodigy, and the first great English evolutionary morphologist. He died in 1882 at the age of 31, but he left a following of several students that did influential work into the next century. In 1880, Balfour asked a question that von Baer could not have answered:

> [Why do animals] undergo in the course of their growth a series of complicated changes, during which they acquire organs which have no function, and which, after remaining visible for a short time, disappear without leaving a trace? (Mayr 1994a: 227, quoting Balfour)

Hypothetical ancestors can be used to explain gill arches and notochords in mammalian embryos. Von Baer's laws cannot. The embryonic organs are not generalized forms of later-developing organs; gill arches are not generalized inner ear bones. Whatever shortcomings the program of evolutionary morphology might have had, they were not caused by the simple failure to read von Baer or to recognize that evolutionary changes can occur in early embryonic stages.

5.4 EARLY ORIGINS IN PHYLOGENY AND ONTOGENY

In this section I discuss examples of evolutionary morphology at work. Each makes use of the biogenetic law. Examples are the germ-layer theory of homology, Haeckel's *Gastrea* theory of metazoan origins, and the complex debates surrounding vertebrate origins.

Germ layers were first discovered around 1820 by Christian Pander, an associate of von Baer. He demonstrated that the bodies of all vertebrates developed out of the same three original layers, which were later named the *ectoderm, mesoderm,* and *endoderm.* One of Huxley's early achievements was the demonstration of the homology between the two inner layers of vertebrates and the two germ layers that make up the bodies of adult coelenterates (jellyfish, corals, etc.). In the 1870s, E. Ray Lankester divided the animal kingdom into grades on the basis of the number of germ layers in the body (one, two, or three). Haeckel immediately adopted the germ-layer origination of a body part as an embryological criterion for homology. This was an extension of the embryological criterion of homology to the earliest stages of embryonic differentiation.

In his most extended exercise in recapitulationist reconstruction, Haeckel went on to hypothesize ancestral forms that corresponded with each major early stage of embryological development in metazoa. The *gastrula* is one important embryological stage. Gastrulation is the event in which the blastula (a simple ball of cells) invaginates to form a cup shape, which then internally differentiates into the two or three germ layers. It is the first point at which an animal begins to have an inside and an outside. According to embryologist Lewis Wolpert, "It is not birth, marriage, or death, but gastrulation, which is truly the most important time in your life" (Wolpert 1991: 12). In a recapitulationist version of the same reasoning, Haeckel inferred that the gastrula must represent an important stage in the evolution of higher animals. It is not merely a stage in the ontogeny of all metazoa; our entire group is descended from an animal whose adult form was the gastrula. The ancestral animal's name is *Gastrea*. Haeckel hypothesized four other earlier ancestors, each corresponding both to a stage in metazoan ontogeny and to a hypothetical ancestor. Some of these ancestral forms have living representatives: A fertilized zygote with a well-formed nucleus corresponds morphologically to an amoeba.

Haeckel's reconstructions of extremely early ancestors were much less bound by morphological data than other debates of the period. Haeckel's eagerness to speculate is part of the reason for his low esteem in the twentieth century. It earned him the epithet *naturphilosophische* from his contemporaries, even though his metaphysical views were quite contrary to his predecessors. Other debates had much more empirical content. These centered on two kinds of questions. One was the origin of a particular character, such as fish fins or vertebrate limbs. The other was the phylogenetic relationship among existing taxa. Both kinds of debate involved the hypothetical reconstruction of ancestors. The example we consider is the origin of vertebrates.[33]

Recall that Geoffroy had attempted to identify aspects of common type between vertebrates and both arthropods and mollusks. Evolutionary morphology traced its emphasis on type to that source. However, the program was committed not merely to identifying signs of commonality but also to tracing the details of common descent. One contender for vertebrate origins was the annelids. Annelids are segmented worms such as earthworms, classified along with arthropods under the phylum Articulata. The other contender was ascidians (sea squirts), which had been regarded as mollusks. Haeckel and Alexander Kowalevsky identified the ascidian connection on the basis of embryology. Anton Dohrn and Carl Semper based the annelid connection on comparative anatomy. In 1866, Kowalevsky published an embryological study of amphioxus, a small eel-like burrowing animal that lacks a spinal column. The animal's overall development was typically vertebrate. It included gill slits, a notochord (an embryological precursor that in vertebrates induces the development of the spine), and neural folds along its back that

---

[33] The narrative is primarily based on Hall (1999a: 84–89). Much more detail is available in Bowler (1996: Chapter 4) and Russell (1916).

fused to form a neural canal. He later discovered that the earliest stages of amphioxus development were not recognizably vertebrate, but they were similar to the early stages of development of the ascidian tadpole. Ascidian tadpoles are active swimmers, but they metamorphose into sessile (stationary) adult sea squirts. They develop with notochords, neural folds, and gill slits, all of which disappear in the adults. Amphioxus and ascidian tadpoles shared the earliest embryological similarities, whereas amphioxus and vertebrates shared later embryological similarities. This appears to indicate that a form like the ascidian tadpole may have been the common ancestor of vertebrates and amphioxus. These homologies were strong enough for Balfour to group amphioxus and the ascidians together with the vertebrates in the new phylum Chordata, to replace Vertebrata.

Dohrn and Semper proposed the annelid theory of vertebrate origins on grounds of comparative anatomy. These anatomists rejected the link from vertebrates to amphioxus and the ascidians, because the latter are unsegmented as adults. Annelids and vertebrates share segmentation, apparently a basic and primitive anatomical trait. We have already seen the importance of segmentation to the morphological tradition in the work of Goethe and Owen. One difficulty with the annelid–vertebrate relationship is shared with Geoffroy's earlier identification of arthropods and vertebrates. It is the relative positioning of the circulatory and nervous systems. Annelids and arthropods have ventral nervous systems whereas vertebrates have dorsal nervous systems. Either the mouth or the brain of vertebrates had to be repositioned during evolution (assuming that vertebrates are the more derived group).

In addition to the flip-over problem for the annelid connection, if segmentation is primitive to the vertebrate–annelid group, what do we say about amphioxus and the ascidians? They show clear embryological affinities to vertebrates, but they lack segmentation as adults. Semper and Dohrn took two different approaches. Dohrn decided that amphioxus and the ascidians are degenerate cyclostomes (jawless fish such as hagfish and lampreys). Cyclostomes were already regarded as degenerate vertebrates who had lost their jaws; amphioxus and the ascidians had further degenerated to the point of losing vertebral segmentation. Semper simply denied that the two were closely related to the vertebrates: He regarded adult segmentation as more indicative of common ancestry than the embryological similarities.

Semper did not reject the importance of embryology, of course. Segmentation itself is an embryological phenomenon. Semper, like Geoffroy before him, regarded the dorsal–ventral distinction to be merely a matter of adaptation, of which side the organism happens to turn toward the sun. The real morphological distinction is neural (nerve side) versus haemal (blood side). Morphological considerations imply that the mouths of annelids and vertebrates could not be homologous structures. They must have arisen independently. Semper recognized that this carries an implication about the common ancestor of annelids and vertebrates: Its process of embryological development must have been subject to modifications that could give rise either to a neural mouth (like annelids) or to a haemal mouth (like vertebrates; see Russell 1916: 282).

This reveals an important principle of the morphological approach to phylogenetic reconstruction. It will come up again. Morphologists conceive of phylogenetic ancestors not in terms of adult organisms but in terms of *ontogenies*. The question is not how one adult form could change into another adult form, but which ancestral ontogeny could be modified to give rise to the ontogenies of descendant organisms. I call the principle that ancestors be reconstructed as ontogenies the *Generative Rule* of phylogenetic reconstruction: Evolutionary transformations are changes in ontogenies, not changes in adult organisms. This principle of scientific practice takes the place of the developmentalist doctrine which was implicit in the work of pre-Darwinian structuralists. The Rule may never have been stated in this form, but it is clearly the shared assumption behind the biogenetic law and other aspects of evolutionary morphology.

A later episode in the debate involved two embryologists who worked through the transition from evolutionary morphology to the research programs that succeeded it. William Bateson worked on the acorn worm *Balanoglossus*. Kowalevsky had discovered gill slits in *Balanoglossus*, and Gegenbaur had proposed on that ground that the worm represented the ancestor to the ascidian larvae and amphioxus. In arguing against the annelid theory, Bateson pointed out that notochords themselves are unsegmented, and that they arise earlier in ontogeny than the (segmented) spinal column itself. Bateson found an apparent homolog to the notochord in *Balanoglossus*, and he argued that this together with the gill slits justified positioning the worm as the closest form to the ancestor of the Chordata. This implied that segmentation must have arisen independently in annelids and chordates. Bateson gave an interesting account of segmentation and one that would lead him in very different directions in his future work. He argued that segmentation merely represented the general tendency of life forms to vary by the spontaneous repetition of existing body parts. Russell points out the similarity of this notion to Owen's concept of vegetative repetition (Russell 1916: 286). Bateson believed that the origin of segmentation required some causal or morphological explanation *other than* common ancestry. This was partly because he disbelieved in the common ancestry of annelids and vertebrates, but it was also because he was coming

117

to doubt the sufficiency of evolutionary morphology to perform the classical task of explaining form. Other embryologists followed on this same path. One who endorsed Bateson's work on *Balanoglossus* was Thomas Hunt Morgan. Bateson and Morgan were among those who turned away from the program of evolutionary morphology and introduced the genetic innovations of the early twentieth century. These are discussed in Chapter 7, but first let us consider the ways in which evolutionary morphology was intended to solve the problem of form.

5.5 EXPLAINING FORM

The phylogenetic explanation of form was the central explanatory goal of evolutionary morphology, just as the phylogenetic explanation of adaptation and diversity is the central explanatory goal of the Evolutionary Synthesis. Evolutionary Synthesis evolutionists do not *reject* the explanation of form – rather, they simply fail to recognize it as a goal of evolutionary biology. This happened for a variety of reasons, many of which are discussed in later chapters. Let us begin with one of the founding documents of the Evolutionary Synthesis.

Theodosius Dobzhansky's *Genetics and the Origin of Species* is at the core of the Synthesis. In that important source, Dobzhansky distinguished two schools of evolutionary biology. One is morphological, and the other genetic (Dobzhansky 1937: 7–8; Dobzhansky 1951: 10–12). The morphological school is said to be interested in historical questions, and the genetic school in causal questions. Dobzhansky reports that most nineteenth-century evolutionary studies were historical–morphological. He explains this by the fact that the idea of evolution was still controversial, and phylogenetic histories provided evidence for common descent. Such evidence was no longer needed in the twentieth century, and interest was turned to the causal question (meaning genetics). Darwin (Dobzhansky reports) was one of the very few nineteenth–century figures interested in causal rather than historical questions: "In this sense genetics rather than evolutionary morphology is the heir to the Darwinian tradition" (Dobzhansky 1937: 8).

Evolutionary theorists lost interest in phylogeny during much of the twentieth century. Dobzhansky sees phylogeny to serve no other purpose than historical narrative. Bowler's book about evolutionary morphology shares this perspective. Reconstructed phylogenies were *Life's Splendid Drama* (Bowler 1996). And they were *mere* drama, with no explanatory purpose. This judgment makes perfect sense in its context, but it misses a crucial aspect of phylogeny as it was practiced in the nineteenth century. Phylogeny was not a mere drama, and it was not mere evidence for past evolution. The research program was addressing causal and explanatory questions, not merely compiling a list of ancestors. Phylogeny had an explanatory purpose.

Consider the analogy to human genealogy. One can study genealogy for merely descriptive or historical purposes, as most genealogical hobbyists do in searching out their ancestors. In contrast, one can study particular genealogies in pursuit of scientific explanations. An example of this is the genetic examination of the detailed genealogies of the nation of Iceland, which may provide knowledge about genetic causes of disease. The evolutionary morphologists' study of phylogeny was not quite like the genetic studies of Icelandic genealogies. The morphologists were *constructing* their phylogenies, not merely reading them from public records. Like the Iceland geneticists, however, they had an explanatory purpose. Their purpose was the explanation of form. Dobzhansky and Bowler do not recognize this goal, because it is not a goal of neo-Darwinism. Nevertheless, the explanation of form was a central explanatory aspect of evolutionary morphology. Let's see how.

As seen in Gegenbaur's statements in Chapter 5, Section 5.2, the commonalities of type were due to the interaction of heredity and adaptation. Adaptation caused divergence, and heredity caused commonality of form. The task of morphology ever since Goethe and von Baer had been to explain form. This includes the origin and transformation of form in ontogeny, and the distribution of form within the Natural System. With the acceptance of evolution, the "relatedness" of similar forms became a literal blood relation. However, the explanatory goal of morphology remains. It is not an option to simply cut out the details, declare that shared forms came from inheritance, and leave it at that. To explain why *this particular* organism has *this particular* form, one must explain both (1) how the form arises in ontogeny, and (2) how that ontogeny arose from ancestral ontogenies during the phylogenetic history of the organism's lineage. Together these explanations supply an account of the relations of form throughout the animal kingdom.

Haeckel's hypothesis of *Gastrea* serves that morphological purpose. Our descent from *Gastrea* explains *why modern metazoa have an inside and an outside.* Wolpert's quip about gastrulation is based on that same recognition: If you hadn't gastrulated as an embryo, you'd have no insides! You may never have asked yourself the question, "Why do I have an inside and an outside?," but the morphologists did. Gastrulation and *Gastrea* are two aspects of the answer – the ontogenetic and phylogenetic aspects.

Consider the brief history in Chapter 5, Section 5.4 of the debate about vertebrate origins. One point at issue was the nature of segmentation in body plans. Did segmentation evolve only once? If it did, then we share our segmentation with earthworms. If it did not, then segmentation is not an evolutionarily unique characteristic, but one that arises in different lineages. If you were to become convinced of the annelid origin of vertebrates, then you share a problem with Dohrn and Geoffroy: Why is the neural side of your own body directed at the sky, rather than directed toward the ground like Articulata? Presumably there was a reversal of dorsal (top) and ventral (bottom) around the time of the origins of vertebrates. I ask you to hold this notion in your mind the next time you look at a lobster (Geoffroy's example). Is it you, or the lobster, which is looking at the world upside down?

I admit that I'm trying to popularize the notion of shared form in this example. We are familiar with popularized adaptationist tales, for example of the similarities between human and bird courtship rituals. With a modified perspective, morphology can be equally intriguing.

Another aspect of the explanation of form involves the identification of the *kind* of cause involved in the origin of a form. The body of amphioxus can be seen in two very different ways. Kowalevsky and Haeckel saw it as representing a relatively unmodified vertebrate ancestor, whose descendents would evolve a segmented spinal column. Dohrn saw it as a degenerate vertebrate, whose close ancestors had lost their spinal segmentation.[34] One cannot know whether amphioxus is a precursor or a simplified offshoot of vertebrates without constructing a phylogeny. Such a phylogeny does not merely name ancestors; it explains the forms of descendants in terms of ancestral forms. Different phylogenies produce different explanations of the forms. The broader question of the phylogenetic frequency of degeneration is a separate morphological issue. Lankester was an embryologist and recapitulationist, but he supported Dohrn's analysis of amphioxus as degenerate. In 1880, he decried the fact that Dohrn's "hypothesis of Degeneration has not been recognized by naturalists generally as an explanation of animal forms" (Lankester 1967: 88). We want the *correct causal explanation* of organic forms, not merely a catalog of ancestors. If degeneration is a frequent event, then we should be very cautious about inferring ancestrality from simplicity. Phylogenies do not explain form all by themselves. Auxiliary hypotheses are needed about the ways that transformation can occur.

We must construct and test hypothetical phylogenies in order to answer the questions of evolutionary morphology. When we do so our purpose is not merely to parade our ancestors, or give one more bit of evidence for common

---

[34] *Degeneration* here means the evolution of a simpler or lower form, not merely a changed form as prefixist transmutationists used the term.

descent. The purpose is to explain form. Brian Hall recognizes this in the program of evolutionary morphology:

> By the late nineteenth century a solution to the generation of organismic form appeared to be at hand in homologous germ layers and conserved stages of embryonic development. This evolutionary embryology was applied to relationships among organisms and in a search for the ancestors of the vertebrates. (Hall 1999a: 69)

I would only add that the explanation of form and the search for vertebrate ancestors were two aspects of the same morphological project: the explanation of form.

This goal was simply not on Dobzhansky's mind when he divided evolutionary studies into historical and causal and then reduced the causal studies to population genetics. By Synthesis convention, "evolutionary causation" refers to population processes such as natural selection *and to nothing else.* The causal processes by which ancestral ontogenies are transformed into those of descendants were so far from Dobzhansky's mind that he couldn't even imagine them as a topic of study.

5.6   THE STRUGGLES OF EVOLUTIONARY MORPHOLOGY

Evolutionary morphology was filled with debates, not only about morphological and ancestral relations, but also about methodological and conceptual issues. One methodological issue was the relative importance of adult morphology versus embryological criteria of relatedness. This had separated Owen from Huxley, and Gegenbaur from Haeckel, and the problem persists to the present day. Another is the nature of causation, and the kinds of causes that were involve in phylogenetic (evolutionary) versus ontogenetic explanations of organismal traits. Haeckel was, of course, the champion of phylogenetic explanation. The first strong argument for the importance of ontogenetic causes came from Wilhelm His in the 1870s. The approach was expanded in the 1880s by Wilhelm Roux and his program of *Entwicklungsmechanik*, later known as developmental mechanics and experimental embryology. His had argued that development should be explained in terms of actual physical "transmitted movement" in the early embryo. This movement folded and rolled the tubes and plates of tissue that made up the embryo. Such physical shaping constituted embryonic development (Gould 1977: 189 ff.; Maienschein 1991a: 45). Haeckel's response to this challenge is puzzling to the modern reader. Haeckel declared that *phylogeny is the mechanical cause of ontogeny* and that His's ontogenetic movements were irrelevant to the explanation of development. One reason given by Haeckel for the irrelevance of ontogenetic "movements" was that, even if these could be found, they would in turn require phylogenetic explanation anyhow. A second was that those movements must be mechanical and deterministic (as long as one was not a teleologist like von Baer). Therefore, unencumbered by teleology, the determinist theorist could skip over the messy details of physiology and treat phylogeny as the cause of ontogeny (Nyhart 1995: 189).[35] In effect, Haeckel was black-boxing ontogenetic causation and then claiming that the "mechanical" (nonteleological) control of phylogeny over ontogeny could be perceived if one ignores the black-boxed ontogenetic interactions.

Haeckel's rejection of the relevance of ontogenetic causation seems absurd today, but it is quite similar to methodological tactics that are well respected. In 1961, Ernst Mayr distinguished between proximate and ultimate causation. Proximate causation included physiological processes such as His's movements, and ultimate causation corresponded to Haeckel's phylogenetic causes. Mayr made the distinction, in part, in order to defend evolutionary biology against the incursions from molecular approaches that were threatening its funding (Beatty 1994). Mayr was a far subtler thinker on this topic than Haeckel, but his reasoning was similar. For example, although bird migrations might be partially explained by the *proximate* genetic causes that control birds' responses to environmental cues, Mayr says that a "complete understanding" requires an *ultimate* explanation of the how the bird got its genes in the first place (Mayr 1961: 1503). Haeckel could have said the same about His's embryonic movements.

Nevertheless, Haeckel's black-boxing of ontogeny was not successful. As the biogenetic law began to show its limitations, the importance of understanding ontogenetic causation became more obvious. Haeckel's anti-ontogeny attitude went against the grain of the morphological tradition. The explanation of form had involved ontogeny ever since von Baer's day, and the biogenetic law was too simple to fit the facts. If it had been true, if terminal addition and condensation really were the only processes by which ontogenies change during evolution, then perhaps Haeckel's attempt to black box the proximate causes of ontogeny might have succeeded. But that is not the world we live in.

Evolutionary morphology was not dependent on Haeckel's extremist version of the biogenetic law, of course, but it did require some way of

---

[35] Although I cannot document it, I have a suspicion that Haeckel's dismissal of ontogenetic causes was his way of undercutting von Baer's arguments that ontogeny must be understood teleologically. If recapitulationist phylogeny were seen as the cause of ontogeny, then von Baerian teleology is refuted.

distinguishing cenogenic from palingenic traits. Cenogenic traits, in turn, required attention to proximate ontogenetic causation. In a striking analogy, Haeckel had likened phylogenetic reconstruction to the deciphering of the original form of an alphabet that had been repeatedly copied, but with letters progressively deleted (compression) or replaced with letters from another alphabet (cenogenesis). Gegenbaur exposed a disanalogy between Haeckel's alphabet and the stages of ontogeny. Ontogenetic stages are not independent units that can be shuffled in and out without affecting other units. A change that happens early in ontogeny can affect the entire later course of ontogenetic development (Nyhart 1995: 249). To understand the effects on form of a cenogenic change, one must have a proximate–causal understanding of the process of ontogeny itself. Haeckel's black box of ontogenetic causes had to be opened.

In the last decade of the century, evolutionary morphology was on the wane (Allen 1978a). Much of the interest turned toward experimental embryology and Roux's developmental mechanics. It is important to recognize, however, that the experimental embryologists themselves had originally intended to contribute to the program of evolutionary morphology. Gegenbaur had shown the importance of ontogenetic causation for understanding the effects of cenogenic change. Roux and others studied ontogenetic causation directly, with the intention that it would lead to better understandings of phylogenetic causation. In the 1894 programmatic introduction of his new journal for developmental mechanics, Roux explained how ontogenetic study would contribute to the understanding of both cenogenic and palingenic change. He pointed out that Gegenbaur's phylogenetic inferences were based on causal assumptions about ontogenetic processes that had not been experimentally verified. Verification of these causes would advance the phylogenetic program (Roux 1986: 133).

In hindsight, evolutionary morphology was trying to do too many things with too few tools. Because of the many exceptions to the biogenetic law, the reconstruction of phylogenetic relationships required knowledge about the relative probabilities of different kinds of changes in ontogenetic processes. (Was segmentation likely or unlikely to arise anew in ontogeny?) That kind of knowledge could have come from two different sources. One is inductive inference from the phylogenies themselves. However, this would be running in a circle, because we can't build reliable phylogenies without prior knowledge of the probabilities of transformations! The other source of knowledge comes from experimental studies of ontogeny itself. This is where Roux comes in. The causal understanding of ontogeny (experimental embryology) will allow us to understand *how ontogenies can change*. Because changes in ontogeny constitute evolutionary changes in body form, experimental embryology can supply us with rules that will allow the construction of phylogenies. These in turn will allow us to explain the evolution of form – or so we would hope.

That was not how it turned out, at least in the nineteenth century. The causal processes of ontogeny turned out to be complex beyond belief. The task of experimental embryology was so difficult that evolutionary morphology was suspended. The major problems of evolutionary morphology began to appear to be irresolvable (at least pending the full understanding of causal embryology). In 1922, William Bateson reminisced about his early career in the 1880s, recalling that "every aspiring zoologist was an embryologist, and the one topic of professional conversation was evolution" (Bateson 1922: 56). However, the program was fading. In 1894, while Roux was hopefully recommending the benefits of experimentalism to the phylogenetic program, others were discouraged. The embryologist Adam Sedgwick, Bateson's former colleague and Balfour's favorite student, pronounced it impossible to distinguish between cenogenic and palingenic traits:

> Embryos of different members in the same group often resemble one another in points in which the adults differ, and differ from one another in points in which the adults resemble; and it is difficult, even if possible, to say whether the differences or the resemblances have the greater zoological value (because we have no clearly defined standard of zoological value). (Sedgwick quoted in Bowler 1996: 80)

The evolutionary morphologists moved in several directions: some toward experimental embryology, some toward the direct study of variation (Bateson), and some eventually toward Mendelian genetics. The century-old goal of morphology, explaining organic form, was fractionated. The ontogenetic study of form flourished during the early twentieth century, with the growth of experimental embryology. However, the link between ontogeny and phylogeny was severed. A new science of heredity was born: genetics. Genetics soon became integral to evolutionary studies by its incorporation into the Evolutionary Synthesis, but genetics itself produced a major gap between embryology and evolutionary biology. In principle, genetics must be involved in embryogenesis. In practice, that involvement was invisible for most of the century. Genetics, *as it was incorporated into the Synthesis*, had not the slightest relevance to the explanation of form. For much of the century, prominent evolutionary theorists saw embryology as irrelevant to evolutionary understanding. Only in the 1990s did the explanation of form again become a central part of evolutionary biology.

5.7   THE CONFLICT BETWEEN ADAPTATION AND STRUCTURE

We have watched the progress of the clash between adaptationism and structuralism from the Cuvier–Geoffroy controversy onward. The adaptationism of Cuvier and the Bridgewater authors gave way to structuralist morphology. For a while, adaptation was in such disrepute that even its discussion was seen as nonscientific. Darwin's *Origin* relegitimized adaptation as a scientific topic, but it did not convince the scientific community of its importance. Evolutionary morphology, a structuralist program, treated adaptation as a side issue. In *Life's Splendid Drama*, Peter Bowler narrates the next episode in this interplay of methodologies (Bowler 1996). He describes how the structuralism of evolutionary morphology was replaced in the new century by adaptationist approaches to phylogenetic reconstruction based on paleontology and biogeography. Phylogenetic studies turned away from typology and toward adaptation and environmental causes of change. This trend did not itself bring about the Evolutionary Synthesis. Discoveries in experimental and population genetics were needed for that. Nevertheless, Bowler convincingly argues that the intellectual climate was favorable to the adaptationism of the Synthesis in the 1930s in a way that it was not in the 1880s.

Bowler describes the trend away from structural explanation and toward adaptation as a move in the direction of truth, not merely in the direction of the Synthesis. He is reluctant even to acknowledge that the morphologists were studying *evolution*. For example, Bowler reports that the evolutionary morphologists did not reject evolution completely; they were only able to incorporate "what they could understand of it" into their accounts, and they found it "difficult to throw off the legacy of the old typological viewpoint" (Bowler 1996: 55, 58). He labels the project of linking ontogeny to phylogeny as developmentalist and historicist, in that it assumes that phylogeny was "equivalent to the embryo's development towards maturity" (Bowler 1996: 16).[36] He describes historical figures whose work is similar to modern adaptationism as taking a "more realistic" approach (Bowler 1996: 101, 258).

Bowler makes two judgments about the objective superiority of adaptationism over structuralism that I believe are mistaken. One is that phylogenetic reconstruction is best conducted by adaptationist scenarios. The other is that the reconstruction of ancestors on the basis of embryological developmental patterns is flawed by its commitment to typology. In each of these judgments, Bowler's views represent Synthesis Historiography and its associated

---

[36] This attribution is particularly unfair. Nyhart shows that this conception was held only by the first two of the six cohorts of nineteenth-century morphologists (see my Chapter 5, Section 5.2).

adaptationism. As an aficionado of the structuralism of modern evo–devo, I see both judgments as having been refuted by the events of the past twenty years.

By the 1990s, the methods of phylogenetic reconstruction came to be dominated by cladistic (also known as phylogenetic) systematics. These methods reject the use of adaptationist scenarios in favor of detailed analyses of the distributions of large numbers of characters. Although they firmly reject adaptationist reasoning, cladist methods also reject the use of embryological reasoning in establishing phylogenies. Thus cladism can be seen as being neutral in the adaptation-versus-structure dispute.[37] Bowler's rejection of the theoretical significance of the ontogeny–phylogeny connection is more serious. I will try to show that this adaptationist critique, if applied to modern biology, would shut down some perfectly legitimate areas of research.

Bowler claims that typologists conceived of ancestral groups as being composed not of "real" organisms, that is, organisms that had real adaptive needs and interacted with a real environment. Instead they conceived of "idealized" embryological processes that produced abstractly described adult morphologies. This practice is said to reveal the pernicious influences of idealist metaphysics and to tie the morphologists to their transcendental precursors. His example of a flawed idealist phylogenist is Adolf Naef:

> Naef proposed a hypothetical developmental pattern which could be modified to produce both annelids and chordates . . . it represented the primitive features of the developmental process stripped of any specialized details that would have had to exist in the ontogeny of a real organism. . . . But since there was no need to think about the starting point as a real organism, there was no incentive to search for the adaptive modifications that might have indicated how the early members of the group had lived, or the kinds of adaptive pressures that might have forced their transformation into the divergent forms we know today. (Bowler 1996: 56)

Even though Naef was writing in the 1920s, Bowler's description of his method is a fair account of the principles used by earlier typologists to unify groups and hypothesize ancestors. It was certainly true of Haeckel's *Gastrea*. None of the reconstructions discussed in Chapter 5, Section 5.4 were concerned with adaptive influences on morphological change, and many of them addressed the ways that ancestral ontogenetic processes could be modified to produce descendant ontogenies.

---

[37] For details on how the cladistic reconstructions of the 1990s repudiate those presented as "realistic" by Bowler, see Amundson (2003).

Bowler's description of the methods of phylogenetic reconstruction are perfectly fair. The only problem with his report is that he mistakenly concludes that the structuralists are "unrealistic," and that only adaptationists are studying "real" organisms. Contrary to Bowler, the difference between structuralists and the later adaptationists is not that one dealt in reality and the other did not. The difference is that they were concerned with different aspects of reality. They followed two different rules of phylogenetic reconstruction.

- *The Generative Rule of Reconstruction*: Identify an ancestral ontogeny that can be modified into the ontogenies of the descendent groups.
- *The Adaptive Rule of Reconstruction:* Identify ancestral characters and selective forces such that the forces might have caused populations that possessed the characters to diverge into the descendent forms.

I reject Bowler's assertion that the Adaptive Rule deals with reality and the Generative Rule with an idealist fantasy world. Ontogenies are real things, and every real metazoan must have one (on pain of remaining a zygote!). Any real organism must follow both rules – it must develop ontogenetically and it must fit its environment. Moreover, any phylogenetic change must *conserve* obedience to both rules – each descendent must maintain both its ontogeny and its fitness. Typologists attended to the Generative Rule and ignored the Adaptive Rule; adaptationists did the opposite. It is no more realistic to ignore ontogeny than it is to ignore adaptation.

My defense of structuralism in phylogenetic reconstruction is justified not only by my nostalgia for evolutionary morphology. It is also justified by whiggish hindsight. The 1990s saw an explosion not only in nonadaptationist cladistic reconstruction but also of the structuralist reconstructions of the body plans (archetypes?) and ontogenetic processes of hypothetical ancestors. A convenient source of comparison is an anthology of papers published during the 1990s in the journal *Nature*, entitled *Shaking the Tree* (Gee 2000). Almost all of the authors reject adaptationist scenarios as tools of phylogenetic reconstruction. One section of the book is composed of evo–devo papers. The similarity to evolutionary morphology and even earlier structuralist biology is not only noticeable to the reader; it is embraced by the authors. The papers commonly discuss developmental processes that are shared between embranchements. In a paper on the dorsoventral body axis, de Robertis and Sasai reproduce Geoffroy's 1822 illustration of a dissected lobster lying on its back to reveal the axis inversion in comparison to a vertebrate. Recent molecular evidence gives dramatic support to Geoffroy's conjecture. Genes designating the dorsoventral axis in chordates are homologous to those in arthropods, but with reversed polarity (De Robertis and Sasai 2000). The authors also allude to Haeckel's *Gastrea* theory. Haeckel does not come out a winner like Geoffroy. *Gastrea* is replaced by *Urbilateria*, the newly hypothesized ancestor of chordates, arthropods, and all other bilaterally symmetrical animals *(Bilateria)*. Other authors discuss the ancestral ontogenies of less-ancient groups. From Sean Carroll's chapter, "It now seems likely that all insect diversity has evolved from a body plan [archetype?] sculpted by the same set of homeotic genes" (Carroll 2000: 76). The chapter by Shubin, Tabin, and Carroll discusses the different developmental roles of homeotic genes in the evolution of vertebrate and arthropod limbs. The limb has been a mainstay of structuralist study ever since Owen's 1849 attack on the adaptationists of his era (Owen 1849). However, even Owen was unwilling to conjecture that developmental similarities between arthropod and vertebrate appendages could have been "retained despite more than 500 million years of independent evolution" (Shubin, Tabin, and Carroll 2000: 100).

Bowler's description of Naef's typological methodology applies perfectly to these writings. Reconstructions yield a hypothetical ancestral ontogeny that could be modified to produce descendent ontogenies. *Urbilateria* is such a construction, representing shared features of the developmental process but stripped of any specialized adaptive details. The reconstruction ignores what Bowler called the "adaptive pressures that might have forced their transformation into the divergent forms we know today." These authors follow the Generative Rule and ignore the Adaptive Rule, as did Naef and the nineteenth-century structuralists both before and after Darwin. Evo–devo reflects the program of explanatory typology. The terms *body plan* and *bauplan* are used with no apparent embarrassment, and even *archetype* springs up regularly. De Robertis and Sasai showed no embarrassment in pointing out the similarities of their views to Geoffroy and Haeckel. Evo–devo revives the old morphological goal of explaining form, a goal that was dormant during most of the twentieth century.

Biologists are again pursuing the problem of form. This is a remarkable development in evolutionary biology. Even while I chide Bowler for being unaware of it, I must admit that Bowler's critiques would have been taken as conclusive for most of the twentieth century, by most evolutionary biologists, *and for very good reasons*! The next phase of our narrative will be to explain what those reasons were, and how they came to be accepted. The breakdown of evolutionary morphology was followed in the early twentieth century by developments in genetics and evolutionary theory that led to the Evolutionary Synthesis. Synthesis theorists eventually came to adopt methodological proscriptions against structuralist explanations of the morphologists' kind. Dobzhansky in 1937 already failed to recognize the problem of form as an aspect of morphology, and he saw morphology as a merely histori-cal study. Bowler accepts this view, and he adds a metaphysical critique of morphology as "idealist" that cannot be found in Dobzhansky. This critique itself is an aspect of the Essentialism Story that has haunted the pages of this book. As Nyhart has demonstrated, only the earliest cohorts of morphol-ogists (the Naturphilosophen) conflated the metaphysical and explanatory aspects of the concept of type. The explanatory concept of type strengthened throughout the century. The modern tradition of Synthesis Historiography has anachronistically identified the metaphysical concept of type as infecting all of nineteenth-century and some of twentieth-century biology.

We must ask first how the study of form became invisible (to evolutionists), and second how it became metaphysical. Roughly speaking, the evolutionary study of form was made invisible as a by-product of the innovations of early genetic theory, and their incorporation into the Evolutionary Synthesis. The study of form became metaphysical as a result of the construction of Synthesis Historiography that began about 1959. We will first look at the history of heredity, and how genetics was born. We will then see how genetics was incorporated into the Evolutionary Synthesis, which gave birth to Synthesis Historiography.

6

Interlude

6.1 TWO NARRATIVES OF THE HISTORY OF
EVOLUTIONARY BIOLOGY

My engagement with Synthesis Historiography as a rhetorical adversary is at an end. SH formed the adversarial background against which the historical narrative of Part I was produced. Part II of my book is not constructed against such a background. It is a straightforward attempt to understand the methodological arguments between structuralists and the Evolutionary Synthesis during the twentieth century. Before beginning the new century, let us summarize and deconstruct the implications of Part I. To some extent, the traditional SH narrative and my revisionist narrative are merely different ways of telling the story of the nineteenth century – yin and yang, the chicken and the egg. I want to acknowledge the ways in which this is true. I also want to point out the ways in which I believe that I have actually corrected the traditional narrative. First, some general differences in how a structuralist and a neo-Darwinian will approach a narrative of nineteenth-century biology.

6.2 ONE THEORY OR TWO?

A central difference is how to interpret Darwin's work. SH commentators will tend to see it as a unified whole, the theory of evolution by natural selection. Structuralist commentators will break Darwin's book into its parts, as did Dov Ospovat. Ospovat was a clear sympathizer with nineteenth-century structuralist biology (although I have no idea whether he was aware of its twentieth-century correlates). He introduced Darwin's own early division of his book into two parts, and I followed that division. The structuralist will

130

think of Darwin's theory in two parts: One is natural selection, and the other is descent with modification, the Tree of Life as a projection of the Natural System. This allows the structuralist wholeheartedly to endorse half of Darwin's theory (common descent), and conduct research on its basis alone, while regarding the other half (natural selection) as a separate issue mostly irrelevant to the structuralist research program. This was the attitude of the evolutionary morphologists, and to date it has been the primary attitude of evo–devo practitioners. Neo-Darwinians, in some contexts, are quite insistent that Darwin's theory is *one theory, not two*. This can be seen in their tendency to label non-Darwinian evolutionary theories as "nonevolutionary." They imply that one does not earn the title of "evolutionist" by believing in common descent alone – one must also believe in natural selection as the primary evolutionary mechanism. "If we suppose that a 'true' evolutionist has to go the whole way in accepting Darwinism, then any biologist retaining a vestige of the idealist or typological way of thinking cannot be an evolutionist" (Bowler 1996: 58). Bowler is strongly tempted in this direction, and he frequently questions whether morphological theorists are "truly evolutionary" even when they are openly committed to descent with modification (see also Ghiselin 1997). This sounds to me like a sectarian dispute about what it takes to be a "true Christian." Then again, I'm a structuralist (so perhaps not a "true evolutionist" by their definition anyhow); of course it sounds that way to me!

The very fact that I stressed the separation between common descent and natural selection in my Part 1 is a commitment to a structuralist rather than a neo-Darwinian interpretation of history. This is not a question of how Darwin viewed his own theory – he went back and forth depending on the context, sometimes insisting on natural selection and other times claiming that common descent was more important. By choosing to depict Darwin's achievement either as unitary or as binary, we modern commentators are choosing a context in which to view history. I chose a structural context. I admit that my choice was not dictated by the facts of history; neither was the choice made by neo-Darwinian commentators.

6.3 GROUNDS FOR SPECIES FIXISM

Synthesis Historiography got this wrong. There is no evidence that essentialist–typological metaphysics was behind a belief in species fixism, and there is plenty of evidence to the contrary. I am not satisfied that we yet know the real grounds for species fixism. SH was so influential that the subject has hardly come up (but see Müller-Wille 1995). It is true that philosophers have located a small number of examples of apparent arguments for species fixism based on metaphysics, and a few openly fixist and special–creationist works that endorse Owen's typological writings. However, these only appear in the 1850s (M'Cosh and Dickie 1855; Dana 1857; Agassiz 1962, first published in 1857). To my knowledge, there is no earlier evidence that metaphysics was responsible for species fixism. Writings of the 1850s are far too late in the game – about a century too late – to show us the real basis of species fixism. They are better understood as last-ditch attempts to avoid the evolutionary implications that were becoming increasingly apparent.

Two questions remain: Why, then, was the Essentialism Story invented in the first place? Why did it persist for so long (if it has so little historical grounding)?

The modern historical context of the origin of the Essentialism Story is discussed in Chapter 10. A simple answer is that essentialism was the most direct and dramatic contrast with population thinking. It was a good way to dramatically introduce population thinking to the general public. However, a second reason was present at the start. When Ernst Mayr originated the story in 1959, he was at least as concerned with twentieth-century non-Darwinian challenges to the Evolutionary Synthesis as he was with pre-Darwinian species fixists. The Essentialism Story aligns the beliefs of twentieth-century non-Darwinians such as Richard Goldschmidt with pre-Darwinian species fixists. By tarring modern non-Darwinians with the same brush as pre-Darwinians species fixists, history could serve a good contemporary cause.

Why did the Essentialism Story persist so long? It does serve a historical purpose in structuring Darwin's achievement. By depicting Darwin's adversaries in a certain way, it depicts Darwin's achievement in a certain way. If his adversaries were species fixists, then his achievement was to prove the malleability of individual species. Natural selection is how Darwin proved the malleability of species, and it is therefore at the center of his achievement.

But what alternative is there? Isn't it obvious that Darwin's primary achievement was the refutation of species fixism? No. Not to a structuralist – at least not the structuralist writing this book.

6.4   DARWIN'S OTHER PRIMARY ACHIEVEMENT – THE TREE OF LIFE

Structuralists are far more interested in the Unity of Type than in the malleability of species. This often earns the scorn of SH commentators. Did

Darwin contribute to knowledge about the Unity of Type? He certainly did! The knowledge that he produced on this topic did not even rely on the truth of natural selection. A structuralist can see an entirely different achievement in the *Origin* than the rejection of species fixism (see my earlier discussion of Gegenbaur in Chapter 5, Section 5.2).

As we saw in Chapter 2, fixism was a new belief in the mid-eighteenth century. Prior to that, a chaotic form of transmutationism was widespread. Wheat gave rise to rye, and barnacles gave rise to geese. Darwin's achievement in 1859 was to demonstrate a different, nonchaotic kind of transmutation. Darwin's theory *explained the structure of the Natural System.* Darwin's transmutationism was *constrained* by the Natural System. Prefixist chaotic transmutationism had explained nothing at all about the Natural System; the Natural System didn't even exist until chaotic transmutationism was replaced by species fixism. Darwin's *Origin* was designed not merely to prove that species can change, but also to prove that *the shape of the Natural System can be explained by common descent.* The mere refutation of species fixism did nothing to demonstrate that the Natural System reflected a genealogy, a Tree of Life. The prefixists had already believed in transmutation, but their opinions were valueless. Darwin's transmutationism was of value *because it was constrained by the Natural System.*

If Darwin's proof of common descent really was an important achievement independent of his proof of species malleability, then who were his adversaries? His adversaries were the taxonomic nominalists – Jussieu, Cuvier, and the Bridgewater authors. For the taxonomic nominalists, the Natural System did not depict a real, objective structure in the world. Darwin's achievement required taxonomic realism. For the Natural System to be a real constraint on a theory of evolution, it had to depict a real structure in the world. This means that Darwin had allies. The taxonomic realists such as MacLeay and Strickland and the idealist morphologists such as von Baer, Barry, and Owen were all allies. They established the objective reality of the Natural System. Darwin showed how to reinterpret the *real* Natural System as a genealogy. He could never have done that if taxonomic realism hadn't been established by others.

The Essentialism Story makes it impossible to recognize this aspect of Darwin's achievement. By focusing all attention on species fixism versus malleability, and none on the Natural System, it degrades the importance of Unity of Type. This is all well and good for neo-Darwinians, of course – the history turns out right for them. However, once we recognize the importance of early structuralist biology for contributing to the reification of the Natural System, we see that structuralist thought was *not antievolutionary at all* (at least in its outcome). Indeed, it was an absolutely necessary contribution to Darwin's achievement. Even the essentialist language used by Strickland and Owen was in support not of a priori metaphysical doctrines, but instead the reality of the Natural System.

6.5 THE SIGNIFICANCE OF GAPPINESS

Neo-Darwinian commentators often regard historical beliefs in continuity between groups as indicators of progressiveness in biological thought. Continuity between groups seems more consistent with Darwinian gradualism in evolution. Gappiness between groups seems (to these authors) to indicate essentialism and therefore scientific regressiveness. A structuralist will consider just the opposite. If there are no gaps, then there are no types, and so there is no reason to construct the grand genealogies that organize organic forms. Here are examples of both kinds of reasoning.

David Hull recently discussed William Whewell's report on the "Type Method." Whewell's Type Method was a report on the systematic methods of Cuvier, which I described in Chapter 2 as the method of exemplary types. The type in this system is arbitrarily chosen, and it is used as a base for describing other members of the taxon. The boundaries between taxa at all levels were conceived to be fuzzy, not discrete. To Hull's credit, he recognized that Whewell's Type Method did not fit neatly with the traditional definition of typology–essentialism. Groups at the edges of one taxon showed similarities with adjacent taxa. Hull found this to be a potentially progressive attitude, because it left open the possibility of transitional forms between higher taxa. Whewell did not actually endorse evolution, but his (Cuvier's) Type Method would have allowed it, and Hull looks favorably on the fuzzy boundaries. "Though fuzzy boundaries are compatible with the gradual evolution of species, they do not necessitate it" (Hull 1999: 58).

I (as a structuralist) am stunned. Here is one of the few times that David Hull looks with favor on a pre-Darwinian author, and (in my view) he picks exactly the wrong time to do it! Whewell is endorsing Cuvier's taxonomic nominalism. The fuzziness – the nongappiness of taxonomic boundaries – means that the Natural System is a continuous smudge of vague similarities. To my mind, this makes common ancestry irrelevant. Without gappy boundaries at the higher taxonomic ranks, there are no grounds for realism about the Natural System. Without a realistic Natural System, Darwin's transmutationism is no better than that of the prefixists. To Hull's mind,

Whewell's Type Method is progressive in that it opens the possibility of gradual and continuous change at all taxonomic levels. Hull concentrates on the continuity entailed by species malleability. I concentrate on the taxonomic gappiness needed for Darwin's Tree of Life. We see with different eyes.

A second example of the way SH authors favor continuity comes from Peter Bowler. Throughout *Splendid Drama*, he criticizes typologists for their reluctance to hypothesize intermediate forms between morphological types (Bowler 1996: 43, 49, etc.). He appears to consider any acceptance of discontinuity to be regressive. Again I disagree, and for the same reason. The *reality* of types and the gappiness of taxonomy were progressive, in that they supported the objective reality of the Natural System. There is plenty of time to search for remote ancestors after the fact of common descent is established. If smooth continuities exist between every taxon and its adjacent neighbor, then we have no reason to believe in an objective *genealogical tree* at all.

Structuralists misinterpret historical sources also, but in the opposite direction. They put evolutionary words into the mouths of nonevolutionary structuralists. Prominent evo–devo researchers Brian Hall and Jessica Bolker have each claimed that von Baer believed that embryology proved something important *about phylogeny* (Hall 1999a: 71; Bolker 2001). But von Baer opposed transmutation, considering it an extravagant hypothesis akin to Naturphilosophie (Nyhart 1995). He did believe that embryology was the key to the Natural System, but he did not interpret the Natural System as a genealogy, and so he disbelieved in phylogeny altogether.

Somehow I find myself less upset about the factual errors of Hall and Bolker than (what I see as) the misinterpretations of Hull and Bowler. After all, the Natural System was transformed by Darwin into the Tree of Life, the phylogenetic pattern. Like Hall and Bolker (but unlike Hull and Bowler), I see Darwin's primary achievement to have been the transformation of the Natural System into the genealogical Tree of Life, not the argument for the malleability of individual species based on natural selection. Thus, I am inclined to be tolerant of Hall and Bolker but indignant toward Hull and Bowler.

Nevertheless, I must admit that the debate is structure versus function, the chicken or the egg. Darwin can be seen as having one theory, or as having two. But I must insist on this: There is both a chicken and an egg. The history of pre-Darwinian biology can be seen in two ways. The Essentialism Story is one way. Part I of this book is another.

6.6   AND FORWARD

Part II of this book will forge ahead into the twentieth century. The first question to be addressed is how development was written out of evolutionary biology. According to the Essentialism Story, it should never have been there – only metaphysical confusion made development seem relevant to evolution. However, we have seen the shortcomings of the Essentialism Story. We are now about to consider how the story came to be told in the first place.

II

Neo-Darwin's Century

Explaining the Absence and the Reappearance of Development in Evolutionary Thought

7

The Invention of Heredity

The biological facts of heredity seem obvious to modern thinkers. We might suppose that the basic facts have been known forever. Among the truisms of heredity are these:

1. Offspring resemble their parents more than they do other members of their species.
2. Heredity is the passing of traits (or representatives of traits, such as genes) between generations.
3. Heredity is primarily a relation between parents and offspring.
4. Hereditary traits (those from our parents) are our deepest and most natural traits.
5. Heredity is independent of development.

None of these truisms is ancient. Truisms 2, 3, and 5 were accepted by the scientific community only in the twentieth century. Truisms 1 and 4 were accepted in the nineteenth. During the eighteenth century, most experts would have rejected all five.

In this chapter I sketch the history of the concept of biological heredity and how we came to accept the truisms. My narrative is mostly progressivist; we have done a good job in discovering the facts. However, I take a "constructivist" stance on at least one issue. In the paragraphs that follow, I will claim that Truisms 2 and 5 are not discoveries but stipulations. At a certain point in history, heredity-theorists stood at a semantic crossroads. Two parties to a theoretical dispute claimed the legitimate ownership of the term *heredity*. With the victory went the semantic spoils: *Heredity* now means what the winners of that theoretical debate took it to mean. The winners were geneticists, and heredity now means genetics. The losers in that debate were embryologists,

139

who considered heredity to be a matter of embryological development. Even though evolution was not a part of the debate, the outcome was momentous for the shaping of evolutionary biology.

I begin the narrative with the earliest discussions of heredity; we will see how it became an aspect of embryology in the nineteenth century. We will then see how heredity was carved away from embryology in the twentieth century. My purpose is not exactly to challenge the modern truism "heredity is genetics," but to understand the history of the relation between evolution and development. Heredity has always been associated with evolution; in the nineteenth century it was also associated with development. Modern views of the relation between evolution and development are strongly affected by the cleavage between heredity and development. Let us see how that took place.

7.2   EPIGENETIC ORIGINS OF HEREDITY

The biological meanings of *heredity* and its cognates are metaphorical, and they are surprisingly recent. The original meaning is the passing on of property or social position between generations. The earliest biological uses of *heredity, inherit,* and *inheritance* in the *Oxford English Dictionary* are all in the mid-nineteenth century, illustrated by quotations from Darwin and Herbert Spencer. The chief biological problem of the eighteenth century was much broader than heredity – the origin of form in the embryo itself. How is generation (reproduction) possible at all? The competitor theories were preformationism and epigenesis. According to preformationism, the earliest embryo already has the form of an adult, and embryological development is merely the unfolding ("evolution") of the already-existing form. According to epigenesis the original embryo is formless, and its form arises during its development. Contrary to modern prejudices, the preformationists were the advocates of mechanistic causation. Most epigenesists of the era advocated special vital forces. This was because everyone agreed that form cannot arise in the egg ex nihilo. Epigenesists claimed that form arose gradually in the embryo, and did so by way of a purposive vital formative power that acted during development. Preformationists rejected purposive and vital forces. Their alternative was to assert that that form already existed in the germ of the egg, having been put there at the time of God's creation in the remote past. Purposive formation had still existed, but only once at the moment of creation. In this way, the preformationists were similar to the liberal theists (Owen, Carpenter, and Baden Powell) in the pre-Darwinian debates: God created, but did so only once at the beginning of the world. Secondary (mechanical) causes took over from there. Preformationism implied that either the mother or the father, but not both, was the repository of the preformed germs. The germs lay like Russian dolls inside the adult parent, each germ nested inside its parent germ. The reproductive role of the non-germ-carrying parent was either to stimulate (if the non-germ-carrier was the father) or to nurture (if the non-germ-carrier was the mother) the preformed germ that was carried by the other parent. The modern disdain for preformationism is quite whiggish. Preformationism was as well motivated as epigenesis, and it was considerably more modern in its metaphysical assumptions.[38]

A moment's reflection will show why preformationists had little interest in resemblances between parents and offspring. The preformed traits of the offspring had been there since creation. There was no theoretical reason to expect offspring to resemble parents. Preformationists could invoke the sharing of nourishment and environment to explain some simple effects, such as the blending of the skin colors of parents in their offspring. However, parental resemblances would be theoretically important only to epigenesists. A proof that significant resemblances exist with both parents would be a serious blow to preformation, because only one parent could have carried the germ. Thus it is not surprising that the first discussion of what we would call "heredity" was for the purpose of refuting preformationism (Jacob 1976: 68). What kinds of traits did the epigenesists cite? Modern thinkers can list dozens of genetic parental resemblances – eye color, hair texture, and dozens of other subtle individual differences. The resemblances noticed by early epigenesists were very different. Pierre-Louis Maupertuis argued for epigenesis in 1745 on the basis of human pedigrees that demonstrated inheritance from both parents. Maupertuis's traits were not subtle ones such as eye color. The doubly inherited traits were polydactyly and albinism. These were dramatic, pathological variants. This illustrates how eighteenth-century heredity was primarily a medical concept, not a concept of natural history (Gayon 2000: 85).

Another empirical claim of the period is even more dramatic. Albrecht von Haller vacillated during his career between epigenesis and preformation. As an epigenesist in 1747, Haller claimed that offspring *but especially hybrids* resemble both of their parents (Roe 1981: 25). His emphasis on hybrids indicates the traits in question: not subtle variations, but traits that potentially differed between species. Haller soon began to doubt epigenesis, and he ended his life as a preformationist. In 1752 he criticized the epigenetic theories of Buffon.

---

[38] On the preformation–epigenesis debates, see Roe (1981), Pinto-Correia (1997), and Sloan (2002).

Haller had already changed his mind about parental resemblances. He admitted that hybrids were intermediate between the species of their parents, but he flatly rejected Buffon's claim that (nonhybrid) offspring resembled their parents more than they do other members of their species! His own research (he said) had demonstrated a great deal of variation among individual humans, with no special resemblance between parents and children (Roe 1981: 28).

As epigenetics gained ground, parental resemblances began to be recognized. Haller's epigeneticist adversary was Kaspar Wolff. Wolff's views came close to those of the nineteenth century. Wolff tried to explain the epigenetic origin of form by the special formative powers of a certain kind of matter, *materia qualificata vegetabilis*, which controlled growth ("vegetation"). The different kinds of such matter would epigenetically produce the structures that characterize different species. The formative matter itself was passed from parents to offspring. Wolff insisted that the form of the offspring was not a copy of the form of the parent. Rather, parental resemblances arose from the fact the bodies of both parent and offspring were formed by the same epigenetic process. Therefore, one does not literally inherit one's parents' traits but rather the formative materia that had epigenetically produced the parents' traits. To account for parent–offspring similarities *in bodies*, the epigenesist appeals to similarities in *the forces that build bodies*. Bodies do not pass from parent to child; epigenetic processes do. Inheritance is the production of parent–offspring similarities, and this production takes place throughout epigenesis. Heredity is an epigenetic process.

This conception of heredity continued into the nineteenth century, even as epigenesis itself changed. Johann Blumenbach was converted to epigenesis around 1780 by observations of regeneration in hydra. His version involved a phenomenally described embryological force. Like Newton's gravity (and unlike Wolff's materia), Blumenbach's phenomenal force made no reference to noumenal causation. Kant himself paid careful attention to the debates, and his regulative–heuristic approach to teleology was influenced by Blumenbach's phenomenal epigenesis (Sloan 2002). Following Kant and Blumenbach, teleomechanists like von Baer differed from the eighteenth-century epigenesists in that they would *heuristically assume*, rather than trying to causally explain, the directedness of the processes of embryogenesis (Roe 1981: 152). Von Baer assumed the epigenetic shaping of the embryo, and he went on to empirically discover the four laws that described the relations between epigenetic processes and adult taxonomy. Epigenetic processes produced all similarities and differences in adults, including those that constituted Unity of Type at its various hierarchical levels. Among the similarities were

within-species similarities and parental resemblances. In other words, heredity was epigenetic.

I briefly discuss three epigenetic accounts of inheritance. The first is
unsurprising: Martin Barry's account of von Baerian embryological thought.
The other two may be unexpected: the epigenetic heredity concepts of Charles
Darwin and August Weismann.

7.3.1   Martin Barry

Recall (from Chapter 3, Section 3.4.1) that Martin Barry's primary concern
was to show that Unity of Type is reflected in the patterns of divergence
during embryological development. Barry claims that this pattern of gradual
divergence and increasing heterogeneity reveals the underlying causes of
embryonic development. The germ of an animal must contain what Barry calls
"innate (plastic) properties" that govern the course of its development. The
plastic properties are derived from the parents. They operate successively and
hierarchically, bringing the embryo through the increasingly heterogeneous
stages, described by von Baer, that reflect Unity of Type:

> If the germ be animal, its leading properties are those characterizing *animals* in
> general. But it has others, common respectively to the class, order, family, genus,
> species, variety, and sex, to which the germ belongs. Lastly, it has properties that
> were previously characteristic of its parent or parents; in which, indeed, all the
> others are included. But no innate properties, except those merely *animal*, are
> at first, to our senses at least, apparent in the structure of the germ. The sum of
> these innate (plastic) properties, determines the direction taken in development;
> determines, therefore, the structure of the new being. (Barry 1837a: 137–138)

Starting from the germ itself, successively more heterogeneous and specific
plastic properties take effect during development, until the particular properties of the organism's parents are produced. Barry quotes von Baer as stating,
"every step in development is possible only through the condition preceding
[it]" (Barry 1837a: 140). Barry believed that all parental traits were conveyed
to offspring, including acquired traits. The causes of *heredity* are exactly
the same as those of *development*. The plastic properties that cause development also cause the similarities between generations, and (for that matter)

conformity to the Unity of Type. Barry's Tree of Animal Development (shown in Chapter 3, Section 3.4.1., Figure 1) depicts heredity in action.

7.3.2 Charles Darwin

The theory of pangenesis was on Darwin's mind throughout his adult life. Unlike Barry's equivocal description of "plastic properties," pangenesis was an openly materialistic and particulate theory. Body parts of parents gave off gemmules, hereditary particles that flowed through the bloodstream and collected in the gonads. Parental gemmules were combined in the new embryo, where they eventually produced body parts that resembled the parts that emitted them in the parents. Darwin considered Unity of Type as a mere by-product of common ancestry, so he made no attempt to embed it in his heredity theory. Nevertheless, Darwin too considered heredity an aspect of development. In a detailed study of Darwin's lifelong views on heredity, Rasmus Winther shows that pangenesis tied together all aspects of variation, heredity, and development. "[H]eredity for Darwin was a developmental, not a transmissional, process. Variation occurred when the environment caused a change in the developmental process of [ontogenetic] change" (Winther 2000: 426). Darwin considered all traits to be heritable; gemmules were produced constantly by all body parts. If body parts became modified, they produced modified gemmules, which were then passed on. Variation was a developmental modification in the parental body, caused by the environment and passed on to offspring.

7.3.3 August Weismann

Weismann holds a special position in the historiography of neo-Darwinian evolutionary theory. Modern neo-Darwinism is opposed to the inheritance of acquired characteristics, that is, so-called Lamarckian inheritance. Darwin himself had not only accepted Lamarckian inheritance but designed his theory of pangenesis to account for it. From the modern perspective, Weismann got Darwin off the hook of Lamarckism. Weismann opposed Lamarckian inheritance, and he produced a heredity theory that prohibited it.[39] Weismann distinguished between two cell lineages in an individual organism: the germ line and the soma. The germ line, made up of the cells that could be contributed to the next generation, was said to be "sequestered" and isolated from the developing body (the soma) of the organism. Weismann described

---

[39] His theory was the first to be labeled "neo-Darwinism" for this reason.

the sequestered germ line as "immortal," passing from generation to generation unchanged. This allied his views with the theories of ancestral heredity of Francis Galton and his followers; the germ line (or Galton's "stirp") links an organism with its entire ancestry, not just its parents. Sequestration also implies that any ontogenetic adaptations that occur in the soma (thicker fur in colder winters, larger muscles in blacksmiths) are not written into the heritable material. Thus Lamarckism is blocked. Some modern authors claim that Weismann's germ–soma distinction proves that understanding development is irrelevant to understanding heredity, and therefore to understanding evolution.[40] However, this is a very modern view. Weismann, like his contemporaries, considered heredity to be very much a developmental matter.

Like Darwin, Weismann held a particulate view of heredity. Unlike Darwin, he did not believe that hereditary particles were produced by adult body parts. Hereditary particles were passed through the germ line. Given Weismann's modern reputation, one might think that the doctrine of sequestration was invented ad hoc to block Lamarckian inheritance, but this is far from obvious. Sequestration served a crucial purpose in Weismann's theory of embryological development, and a purpose that was quite independent of Lamarckism. This particular integration of heredity with development deserves examination.

The central problem of the study of embryological development is explaining the increase in heterogeneity in the developing embryo. Seen in terms of the cell theory, increased heterogeneity could be conceived as cellular differentiation. How does the single cell of the zygote give rise through division to the specialized cells of the various parts of the body? The answer given by Weismann and Roux was the *mosaic theory* of development. Roux stated his version of the mosaic theory in 1885, the same year that Weismann proposed the germ–soma distinction. Mosaic or autonomous theories of development assert that the nature of body parts is determined in advance of their actual development, and determined independently of the body parts around them. In contrast, regulative theories of development claim that body parts take their nature from their position within the embryo.

The Weismann–Roux mosaic theory explains the differentiation of the embryo as a direct consequence of differentiation within the germ plasm that is carried within somatic cells. The first somatic cell contains in its germ plasm all of the determinates for the entire body. Cell divisions divide the germ plasm and pass unequal portions to each daughter cell. The first division might distribute the determinates of the right and left halves of the body into the

---

[40] These arguments are discussed in Chapter 11.

two resulting daughter cells. Subsequent divisions break down the hereditary material further (say, dorsal and ventral) and so on, until the single hereditary particles are reached that determine the nature of fully differentiated somatic cells. Cells that receive only bone-determinates become bone, those that receive only skin-determinates become skin, and so on. Adult somatic cells contain only those determinates that specify their particular cellular nature. So the differentiation of the body is caused by the parallel differentiation of the germ plasm contained in the cells. Both differentiations take place during cellular division.

At this point a question arises: If Weismann's theory of mosaic development is true, how is reproduction possible? Adult somatic cells have had their hereditary determinates reduced to a minimum – they are not competent to pass on the determinates of other kinds of differentiated cells. There is only one possible answer. The germ line *must be sequestered* prior to the somatic cell divisions. Reproduction of offspring cannot come from the somatic line, because no somatic cell has a full complement of hereditary determinates. (If it had, it would not have differentiated in the first place!) Without a sequestered germ line, reproduction would be impossible. This follows from the mosaic theory of development alone, *without regard for Lamarckian inheritance*. Which factor was more important to Weismann, his mosaic embryology or his anti-Lamarckism? I cannot judge. The large secondary literature concentrates on Weismann's views on the germ line and Lamarckism, not on somatic development. Even so, recent scholarship affirms that Weismann considered heredity to be an aspect of development (Griesemer and Wimsatt 1989; Griesemer 2000; Winther 2001). The fact that germ line sequestration was tightly integrated within both his evolutionary and his embryological thought merely reinforces this point.

It is important to recognize what is and what is not explained in the Weismann–Roux embryological theory. Differentiation among body parts is explained as a consequence of unequal distribution of determinates during cell division, but the particular characteristics of the differentiated cells are not explained. The mosaic–particulate theory explains why differentiation occurs at all, but not how the hereditary determinates caused the properties that they were responsible for. The units of development–heredity were undefined except by the adult traits that they were postulated to explain. Bone cells are bone because they have bone-determinates; that is all we can say. (Opium puts people to sleep because it has the soporific power.)

This explanatory gap did not go unnoticed. Following Weismann's publication of his full theory, Oscar Hertwig wrote a detailed critique, *The Biological Problem of To-Day: Preformation or Epigenesis?* (Hertwig 1894). Even though Hertwig agreed with Weismann that heredity was localized in the cell nucleus, he criticized Weismann's theory as preformationist. Weismann explained properties of the observed embryo by assigning them to corresponding properties of the unobserved germ plasm. Such theories give the impression of causal explanation where none truly exists:

> When, to satisfy our craving for causality, biologists transform the visible complexity of the adult organism into a latent complexity of the germ, and try to express this by imaginary tokens ... they prepare for our craving a slumbrous pillow. ... [Weismann's method] transfers to an invisible region the solution of a problem that we are trying to solve, at least partially, by investigation of visible characters. (Hertwig 1894: 11, 140)

Hertwig's opposition was substantive as well as methodological. He proposed a regulative, epigenetic account to replace Weismann's mosaic account of development. Cells became differentiated as a result of interactions with other cells of the developing embryo. Development could only be understood by studying these interactions, not by hypothesizing preformationist heritable particles defined only by their effects on the adult.

The epithet *preformationist* was to persist in later debates, and it lives on today. To call a theory preformationist is to claim that the theory *assumes as a given* some aspect of a developmental phenomenon that the critic believes should receive a causal, developmental analysis (Maienschein 1999). Weismann and Roux were not preformationists in the literal sense of the eighteenth century, of course. Old-style preformationism held that form itself preexisted in the embryo. Entwicklungsmechanik was the study of *epigenetic* causes, after all. Weismann and Roux were epigeneticists with respect to the parceling out of the hereditary determinates throughout the soma. However, their epigenetic strategy ended with the postulated powers of the hereditary particles to control traits of body parts. Because no specific traits could be identified as being caused by specific hereditary particles, the theory looked speculative as well as preformationist. Many embryologists treated it with skepticism.

The heredity theories of Barry, Darwin, and Weismann differed greatly, but each saw heredity as an aspect of development. They were not alone. One study has concluded that each of thirty distinct nineteenth-century heredity theories was developmental in nature (Sandler and Sandler 1985: 65). I know of only two possible exceptions to this generalization. One is Mendel (although interpreting Mendel's true intentions is immensely difficult; see Sapp 1990). The other is the biometrical work of Karl Pearson. Pearson was a radical phenomenalist and positivist in the tradition of Bishop Berkeley and Ernst

Mach (Pearson 1892; Provine 1971: Chapter 2). He devised statistical measures, and he studied the correlations of traits among generations while disdaining any physiological interpretation of how those traits might be conveyed between generations (or any other questions of physical causation). His epistemological extremism may have had ironic repercussions, which we will examine later. With these two possible exceptions, the epigenetic concept of heredity survived into the twentieth century. It was soon challenged.

7.4   THE CLEAVAGE BETWEEN HEREDITY AND DEVELOPMENT

As we saw in Chapter 6, Section 6.4, Thomas Hunt Morgan was among those who began professional life as an evolutionary morphologist and embryologist, but who abandoned the phylogenetic goals of evolutionary morphology and turned to the study of proximate–causal embryology. Much of Morgan's early work was in regeneration, a model system of epigenetic heredity ever since Blumenbach. He and other embryologists in the early twentieth century still regarded heredity as an aspect of development. In 1908 Morgan's friend E. G. Conklin stated the position clearly:

> Indeed, heredity is not a peculiar or unique principle for it is only similarity of growth and differentiation in successive generations. . . . The causes of heredity are thus reduced to the causes of successive differentiation of development, and the mechanism of heredity is merely the mechanism of differentiation. (Conklin 1908: 90)

Morgan concurred: "We have come to look at the problem of heredity as identical with the problem of development" (Morgan 1910: 449). Similarities in adult forms of parent and offspring are caused by similarities in the patterns of differentiation. Heredity is the passing on of developmental processes.

The particulate, chromosomal theory of heredity had gained acceptance by the turn of the century, and it was enlivened by the rediscovery of Mendel's laws. Like many embryologists, Morgan had rejected Mendelism along with other particulate theories as preformationist. Ontogenetic development was a causal process that resulted in gradually increasing complexity, and such a process could not be mapped onto a sequence of particles that were claimed to be associated with adult traits.

Morgan would soon reverse his views on Mendelism, chromosomes, and the nature of heredity. He was one of experimentalists who began to use *Drosophila* between 1900 and 1910, and his famous "fly room" at Columbia University was an important research center through which passed many of the most important geneticists and evolutionists of the twentieth century (Kohler 1994). Although many of his students would be educated in Mendelism and the chromosome theory from the start, Morgan had been trained as an embryologist. As Scott Gilbert has shown, Morgan's conversion to Mendelism came by an entirely embryological route, in the study of the developmental causes of sex determination (Gilbert 1978). Despite his antipreformationist skepticism, he gradually became convinced both of the chromosome theory and of Mendelism by the discovery of *Drosophila* mutations that segregated with the sex chromosome.

Early in the century, various views existed on the relations between Mendelian "factors" (later genes) and characters. The simplest was the concept of the "unit character" of Hugo De Vries and William Bateson. In this version each factor correlates with one character; indeed the factor and the character often seemed to be conflated. This simplification was difficult to avoid, and it still springs up in popular literature. However, no self-respecting embryologist could imagine an ontogenetic system of such degenerate simplicity. As Morgan became convinced of Mendelism, he insisted on a many–many relation between factors and characters, with some genetic factors affecting many characters (a relation later called *pleieotropy*) and each character being affected by many genetic factors.

In 1915, Morgan coauthored *The Mechanism of Mendelian Heredity* with three of his students. The book became the primary influence in the success of the Mendelian–chromosomal theory of heredity, commonly referred to as the MCTH (see Brush 2002, who suggests that the M may as well refer to Morgan as to Mendel). The final chapter of the book introduces what has come to be called the *differential concept* of the gene (Schwartz 2000). Characters are affected by many factors; twenty-five factors at different loci had been discovered to affect red eye color in *Drosophila*. When a factor at one particular locus mutates to a particular different form, the result is a pink eye color. In this situation, even though all of the twenty-four other nonmutated factors are still affecting the color of the eye, the mutated locus is referred to as *the cause* of the pink eye color.

> In this sense we may say that a particular factor (p) is the cause of pink, for *we use cause here in the sense in which science always uses this expression,* namely, to mean that a particular system differs from another system only in one special factor. (Morgan et al. 1915: 209; emphasis added)

It is easy to overlook what a radical claim this is. Prior to this assertion, "the cause" of any adult body characteristic could potentially include the entire embryological history of the organism, at least from an embryologist's viewpoint. However, if a single allele can be regarded as the cause of pink eye color, then it is possible to causally explain adult characteristics without any reference to the embryological processes that actually brought them about. The authors are well aware that their assertion of genetic causation has cut ontogenetic development out of the explanatory picture.

> The cause of the differentiation of the cells of the embryo is not explained on the factorial hypothesis of heredity.... [Factors are conceived as chemical materials.] The characters of the organism are far removed, in all likelihood, from these materials.... [We can analyze genetic causation] quite irrespective of what development does so long as development is orderly.... Although Mendel's law does not explain the phenomena of development, and does not pretend to explain them, it stands as a scientific explanation of heredity, because it fulfils all of the requirements of any causal explanation. (Morgan et al. 1915: 226–227)

These passages have a kind of Humean, positivist concept of causation that is difficult to account for (although I will make an attempt shortly). Causation is reduced to patterns of conjunction, even though one of the conjuncts is an inferred particle. The detailed mechanical–causal connections between causes and effects, the ideal of Entwicklungsmechanik, are ignored. Given this new quasi-positivist ontological stance, the debate between Weismann and Hertwig is suddenly irrelevant. It doesn't matter whether ontogeny is regulative (epigenetic) or mosaic (preformationist). The fact that correlations can be traced between the *end products* of ontogeny in successive generations (the traits of parents and offspring) is enough to declare that the causes of these end products have been found – whether those end products arose by preformation, epigenesis, or magic. Development doesn't matter to heredity.

Morgan finalized the split between genetics and embryology in his 1926 book, *Theory of the Gene*. Needless criticism of genetics, he said, had come from confusing the problems of genetics with those of development. (Much of that criticism had been his own, of course.) As he stated at that time, "the theory of the gene is justified without attempting to explain the nature of the causal processes that connect the gene and the characters." Once the distinction between genetics and development is recognized, as Morgan later pointed out, we can see that "the sorting out of characters in successive generations can be explained at present without reference to the way in which the gene affects the developmental process" (Morgan 1926a: 26–27). This expression of the relation between genes and traits is more cautiously stated than the

1915 statement.[41] Rather than genes causing characters, the gene theory is said to explain *the sorting of* characters. Morgan's definition of "the theory of the gene" is even vaguer in its statement of the relation between genes and characters. Characters are only mentioned in the first phrase of the definition, with the remainder giving details about the inferred behavior of genes: "The theory states that the characters of the individual are *referable to* paired elements (genes) in the germinal material" (Morgan 1926a: 25; emphasis added). No matter how the gene–character relation was expressed, one point was repeatedly stressed: Genetics explained characters (or the sorting of characters), and it did so in a way that required no attention to development.

Morgan eventually distinguished between two forms of genetics. One was transmission genetics, the Mendelian study. The other was developmental genetics, the study of the physiological action of genes in embryogenesis. The theory of the gene in the sense of his 1926 book was transmission genetics alone, and this became the common usage. Morgan derived two crucial points from this distinction. One was that *heredity is transmission genetics*. The second was that embryologists ought to turn their attention to developmental genetics.

Morgan's cleavage of heredity from development was spectacularly successful. Neo-Darwinians and most historians writing prior to 1980 treat the Mendelian nature of heredity as a simple discovery, like the discovery of DNA as the chemical nature of the genetic material. However, at least one aspect is clearly a convention, not a discovery. Even if we assume the legitimacy of the distinction between transmission genetics and developmental genetics (which many embryologists did not), we must make a decision about which new field takes possession of the term *heredity*. It certainly was a discovery that many traits followed Mendel's laws. The distribution across generations of these Mendelizing traits could be studied in a way that ignored their ontogenetic development. However, it was not a discovery that Mendelizing traits deserved the title of *heredity*. It was a semantic decision, and a contentious one at that. Many embryologists resisted the co-option of the term *heredity*, as I discuss later. Even more of them refused Morgan's recommendation that they turn their studies to the expression of Mendelian genes in development. Furthermore, the implicit claim of the Mendelians that *all* hereditary similarities were carried by Mendelian genes remained controversial for decades . . . and

---

[41] It is possible that the 1915 statement about factors causing characters was written by Muller, who had a more reductionist concept of the relation between genes and characters than had Morgan himself. I owe this point to Raphael Falk and Gar Allen.

possibly even to the present day. Nevertheless, the Mendelians won the semantic battle. Heredity, by the 1930s, was a matter of transmission genetics.
The organism was split into two parts, which came to be called the *genotype* and the *phenotype*, corresponding to the gene and the eye color that it
"caused." The embryological process that connected these two parts was left
unnamed.

As we discussed in Chapter 1, the present era is epistemologically liberal
and nonpositivist. We love our deep theories and our hypothetical explanations. Perhaps my labeling of Morgan's methodology as quasi-positivist will
seem to the reader to be criticism of Morgan. This would misinterpret my
intentions. Positivist and nominalist antirealism plays an important role in
science, especially at times of radical theory changes. Newton's hypotheses
non fingo served to buffer his theory of universal gravitation from the widely
held view that causal interactions could only occur by direct contact – no
action at a distance. Morgan made very much the same move. He claimed
that a hereditary determinate in a zygote could be said to "cause" an adult eye
color, irrespective of our complete lack of knowledge of the developmental
processes by which that causation took place. This is *precisely* action at a distance. Morgan's theory was not fully positivistic, of course – he hypothesized
the physical location of genes on chromosomes, and he encouraged the study
of developmental genetics. Neither of these stances is positivistic about gene
action. Nevertheless, the gene–trait relationship as depicted in transmission
genetics was just as miraculous as Newton's force of gravity – and it was
defended by the same methodological stratagem.

7.5   REINFORCING THE DICHOTOMY: REWRITING WEISMANN AND JOHANNSEN

The MCTH originated a dichotomous view of the organism that has become
so widespread as to seem almost tautologous today. Nothing could be clearer
to the biological thinker of today than the genotype–phenotype distinction.
Nothing could be more obscure during the period around 1910. My concern
in this section is not to challenge the genotype–phenotype dichotomy. It is
rather to point out how different it was from earlier ways of thinking about
organisms, and how problematic our histories are as a consequence.

The genotype–phenotype distinction itself is universally attributed to a
1909 publication by Wilhelm Johannsen. It is often acknowledged that
Johannsen held unusual views; for example, he was opposed to particulate
theories of inheritance even though he coined the term *gene*. It is less often

recognized that his phenotype–genotype distinction meant something very different from the modern use. The terms were very obscurely defined, but it is clear that genotypes and phenotypes were characteristics of populations or lineages, and not of individual organisms (Churchill 1974). The *phenotype* was a statistical description of the "appearance" of characters across a population or within a lineage. The term *genotype* was an abstraction Johannsen introduced specifically in order to discourage speculation about individual particulate genes. It applied most clearly to "pure lines." These were inbred and genetically identical lineages. The so-called genotype of the pure lines had no variation, whereas that of natural populations was often a mixture of distinct pure lines. Johannsen did not refer to genotypes or phenotypes of individual organisms. Only late in his life, after the MCTH had been introduced, did Johannsen reluctantly refer to the genotypes and phenotypes of individuals (Churchill 1974: 24).

The modern individualistic version of the genotype–phenotype distinction has extraordinary power over our imaginations. Churchill's 1974 paper on Johannsen was a real landmark. Prior to its publication, some of our best historians had mistakenly reported that Johannsen distinguished between the genotypes and phenotypes of *individual organisms* (Allen 1966: 53, cited in Churchill 1974: 17; also see Provine 1971: 99). It is frequently stated that Morgan's adoption of the MCTH was influenced by Johannsen's distinction; the two had met in a 1910 conference devoted to Johannsen's pure line studies. However, when we recognize the anti-individualistic nature of Johannsen's distinction, such influences sound suspiciously modern – as if Morgan had spontaneously extracted the *modern individualistic* genotype–phenotype distinction out of Johannsen's obscure populational definitions. The only report on the Johannsen–Morgan relation that I find helpful is by Raphael Falk (Falk 2000: 321). Falk points out that Johannsen had been strongly influenced by Karl Pearson's phenomenalist positivism. Johannsen's positivism may have given Morgan an excuse to overlook the preformationism of the particulate Mendelian theory. This would account for the doctrines just described as quasi-positivist, which said that a hypothesized gene is the cause of a correlated trait "in the sense in which science always uses this expression," even though ontogenetic causality has been expressly ignored. The modern version of the genotype–phenotype distinction is implicit in Morgan's 1915 book, but it did not come directly from Johannsen.

Johannsen and Weismann are frequently presented as the originators of the modern dichotomous view of organisms. When Ernst Mayr points out Johannsen's regrettably typological concepts, he states that Weismann's notions of germ line and soma are actually closer to the modern meanings of genotype and phenotype than are Johannsen's original definitions (Mayr 1982: 782). This, too, distorts history. We can recognize the distortion by considering the reputation of Weismann at the time of the formation of the MCTH.

Recall the significance of the germ line–soma distinction for Weismann's mosaic account of embryological development. Somatic development took place by the unequal distribution of hereditary determinates during cell division. The germ line had to be separated from the soma, because, if it were not, no cell in the body would be able to carry the full set of determinates into the next generation. The separated germ line blocked Lamarckian inheritance. This theory was met with a piece of bad news around the turn of the century. The evidence was mounting that somatic cell divisions were *not* unequal, and that the same genetic material existed in virtually all somatic cells as in germ-line cells. Weismann's argument for particulate inheritance had been coordinated with an embryological theory based on unequal division of determinates. But unequal cellular division had now been refuted! How could Weismann's views be maintained?

In fact, they could not. Weismann was not regarded as the great forward thinker during the first two decades of the twentieth century that he was after the Evolutionary Synthesis was established. Johannsen condemned Weismann's "speculative" theorizing and preferred Pearson's positivism. Morgan discussed Weismann along with Herbert Spencer and Darwin in a brief section of *Theory of the Gene* about historical particulate theories. He reported that Weismann's particulate theory of the isolated germ plasm had been used both to oppose Lamarckism and to explain development:

> The application of his theory to embryonic development lies outside the modern theory of heredity that either ignores the developmental process, or else postulates a view exactly the opposite of that of Weismann, namely, that in every cell of the body the entire hereditary complex is present. (Morgan 1926a: 30)

This "modern theory of heredity that . . . ignores the developmental process" was the MCTH. It was only a decade old, and its dissociation from development (and especially from Weismann's mosaic theory) had to be stated again and again.

A year later, Weismann received praise from a surprising source, an embryologist who rejected the MCTH. Frank Lillie praised Weismann not for *separating* heredity from development but for *uniting* the two. Lillie opposed the separation, and he approved Weismann's view that "the theory of development included the theory of heredity" (Lillie 1927: 361).

The lionization of Weismann for having separated heredity from development would have to wait until later in the century. Then his embryological theories were forgotten, and the germ line–soma distinction could be reinterpreted into a mere statement about inheritance. Prior to the MCTH, there was no such thing as a *mere* statement about inheritance: Every statement about inheritance was simultaneously a statement about development.

This fact itself makes the MCTH an even greater achievement, of course. The theory was not a mere generalization of the germ line–soma distinction, because Weismann's distinction was itself an embryological one, connecting heredity to development by means of unequal cell divisions. The MCTH was not a mere application of the genotype–phenotype distinction, because Johannsen did not distinguish between genotypes and phenotypes of individual organisms until after the MCTH had already done so ahead of him. Morgan managed to convince his audience of a particulate theory of heredity *while admitting* that he had no theory of development to wed it to. This is why I remain intrigued with the quasi-positivist causation intimated in *The Mechanism of Mendelian Heredity*. Morgan was an extraordinarily flexible thinker, as his biographer Garland Allen stresses (Allen 1978b). Morgan had already metamorphosed from an evolutionary morphologist to an experimental embryologist, and now he was becoming a geneticist. Each transition had required the abandonment of previous methodological goals: first the goal of understanding phylogeny through embryology, and second the goal of understanding heredity through development. In successively abandoning his previous research goals, he became one of the most important innovators of the modern view of heredity. His new nonepigenetic concept of heredity delineated a tremendously fruitful field of study. It could be argued, however, that Morgan's new concept of heredity made it appear as though ontogeny was irrelevant to phylogeny by subterfuge: It made ontogeny irrelevant to phylogeny by redefining heredity to exclude ontogeny.

7.6 BROAD AND NARROW HEREDITY

The theoretical importance of heredity, as parental resemblance, was discovered by epigenesists and used to refute preformationists. It was broadened in the nineteenth century to apply to remote and even phylogenetic ancestors, but heredity primarily acted through the processes of ontogenetic development. It may sound paradoxical that heredity was conceived both as ancestral (connecting individuals with their remote ancestors) and as developmental (manifesting itself during embryogenesis). The paradox is easily resolved.

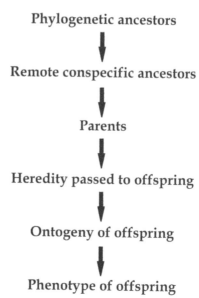

Phylogenetic ancestors

Remote conspecific ancestors

Parents

Heredity passed to offspring

Ontogeny of offspring

Phenotype of offspring

Fig. 5. Broad heredity: Heredity as conceived during the nineteenth century. It connects an organism to its remote ancestors, or (prior to Darwin) to other organisms of its type. Heredity was expressed progressively throughout embryological development. Compare Barry's tree of embryological development (Chapter 3, Section 3.4.1., Figure 1). The earlier stages of ontogenetic development reflect heredity that is shared with remote ancestors, or (prior to Darwin) with higher taxonomic groups, such the animal type and the vertebrate type.

Very early embryological traits, such as gill slits, are shared not only with our parents but with all vertebrates. It makes perfect sense to consider them as "inherited" from our fish ancestors. The individual differences that we inherit *only* from our parents appear late in development. Our gradual embryological development reveals our hereditary linkage to a whole lineage of ancestors, as Martin Barry's description illustrates. One need not be a strict recapitulationist to see a general reflection of phylogeny in ontogeny. This reflection *is* heredity in its nineteenth-century meaning. It is both ancestral and developmental. This was "broad heredity," as shown in Figure 5.

Heredity was drastically narrowed in the early twentieth century, under the combined influences of Mendelism, the chromosome theory, and a smidgeon of positivism. The new concept of "narrow heredity" was a relation among so-called phenotypic traits of subsequent generations, mediated by a hypothetical entity called a gene. Heredity no longer connected an organism to remote ancestors, nor did it refer to the processes of ontogenetic development of traits within an embryo. These embryonic processes must exist, and they

Parents

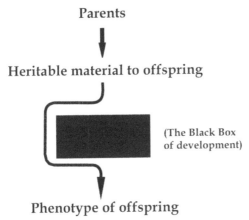

Heritable material to offspring

(The Black Box
of development)

Phenotype of offspring

Fig. 6. Narrow heredity: Heredity as conceived after the inauguration of the MCTH in 1915. Neither ancestry nor development is reflected in heredity. The exclusion of both ontogeny and phylogeny from the concept of heredity may have been the single most important cause of the absence of development from the Evolutionary Synthesis.

must be regular and lawlike, but they were semantically screened from participation in heredity by the concept of hereditary causation associated with the MCTH. Recall how Geoffroy was forced to admit that his identification of homologous bones in distantly related organisms was an "idealization." I suggest that the definition of heredity whereby a hypothetical particle is said to cause a trait *without reference to its ontogeny* is every bit as idealistic as Geoffroy's homologies. Nevertheless, methodology aside, narrow heredity led to an explosive growth of knowledge in genetics. It also led to the separation of heredity from developmental biology, as shown in Figure 6.

Let us reconsider our list of truisms from Section 7.1.

1. *Offspring resemble their parents more than they do other members of their species.* This was specifically rejected by some eighteenth-century preformationists such as Haller.

2. *Heredity is the passing of traits (or representatives of traits, such as genes) between generations.* Epigenetic concepts of heredity generally reject this view. Heredity is the passing of developmental processes, not traits, from parent to offspring. Offspring inherit their modes of development; these modes of development in turn produce the traits.

3. *Heredity is primarily a relation between parents and offspring.* This view was rejected by nineteenth-century advocates of ancestral heredity, including such important thinkers as Galton and Pearson. Preformationists would (trivially) reject it as well.

157

4. *Hereditary traits (those from our parents) are our deepest and most natural traits.* Preformationists considered the traits by which children resemble parents (such as skin shade) to be trivial variations on preformed germs. We will soon see a modern version of this old view: Many critics of the MCTH considered Mendelian traits to be minor variations on the more "fundamental" traits that are conveyed by a separate system of inheritance.

5. *Heredity is independent of development.* As this chapter has shown, this was a very hard-won truism of the twentieth century. The truism became central to the Evolutionary Synthesis, but it was resisted by many embryologists well into the century. Let us turn to that story.

8.2   THE STRUGGLES OF NATURAL SELECTION

We have discussed the fact that natural selection was not considered to be the primary mechanism of evolutionary change by most evolutionists immediately after Darwin. "The Eclipse of Darwinism" was a chapter title in Julian Huxley's *Evolution: The Modern Synthesis* (Huxley 1942). Peter Bowler extensively studied that eclipse, as we saw in Chapter 5. Bowler attributes the unpopularity of natural selection to the biases from preevolutionary thought forms. The influences of typology and idealism, together with an abhorrence for nonpurposive and random mechanisms, made natural selection unacceptable. Jean Gayon has traced the fortunes of natural selection during this period and come up with a very different explanation for its lack of support (Gayon 1998). According to Gayon, natural selection itself was seriously lacking in scientific credentials during the period.

Gayon distinguishes between the heuristic or explanatory use of natural selection and its status as a scientifically confirmed mechanism. There was little question of its heuristic usefulness in stimulating research, or in its explanatory powers. It does indeed explain large bodies of diverse facts, for example in embryology and taxonomy, as Darwin claimed it did. However, natural selection could not be treated as a basic axiom of science, a brute fact of nature like Newton's gravitational force. This is because natural selection (unlike gravitation) was alleged to be a consequence of a set of other, more basic, facts about the operations of nature. These included facts about natural variation, superfecundity, and especially heredity. No matter how richly explanatory and heuristically fertile the notion of natural selection might be, certain facts about heredity would destroy it. Gayon shows two things about the knowledge of heredity during this period. First, the known facts of heredity did not seem to support the operation of natural selection as a long-term cause of continuous evolutionary change. Second, and what is more important, it was never clearly understood exactly what hereditary facts *would* support it! As richly explanatory as natural selection was, no one was able to say exactly what facts about heredity would allow natural selection to operate as the primary cause of evolutionary change. "[I]t was only in the 1920s that the hypothesis of natural selection . . . took on even a semblance of validity" (Gayon 1998: 397).

Gayon discusses several intriguing aspects of heredity as it was conceived by nineteenth-century figures. One is the synonymy of heredity with ancestry and genealogy: Darwin summarized his theory both as descent with modification and as inheritance with modification, and the concept of ancestral heredity was common after Galton. Another is Darwin's own deep equivocation in the *Origin* on whether heredity was continuous or discontinuous. Given the concepts available at the time, if heredity was continuous, then it was blending, and selection couldn't take hold. If it was discontinuous, then selection could take hold, but then the discontinuous determinates that were selected had not themselves been shaped by selection. This was the topic of Fleeming Jenkin's well-known critique of Darwin, a critique that Darwin acknowledged but never successfully dealt with (Hull 1973: 303; Gayon 1998: 85 ff.). A third is the fact that both adaptation and heredity were broadly conceived as *forces* during those times, analogous perhaps to gravity, and working in opposition to each other. This conception was to be modified in an interesting way by the Evolutionary Synthesis.

Gayon looked at nineteenth-century heredity from the Synthesis standpoint, and he noticed aspects of the heredity theories that were problematic for the Synthesis itself. He did not notice the fact discussed in Chapter 7, that nineteenth-century heredity was epigenetic, intertwined with development. Like Provine (1971), Gayon stresses the importance of population genetics in the Evolutionary Synthesis. This is undoubtedly correct, but for population genetics to be the foundation of an evolutionary theory, it must be able to take account of the two general factors involved in evolutionary change: heredity and adaptation. The MCTH concept of narrow heredity allowed this to happen. Narrow heredity divorced ontogenetic development from heredity itself. It associated particulate genes directly with phenotypic traits – traits that can have selective values – and named this a causal connection. This narrow gene–trait relationship *became* heredity, and it allowed the invention of population genetics. Heredity is genes, and adaptation is the result of the Darwinian sorting of the genes by the fitness of their phenotypic correlates. Who cares how complex the ontogenetic route by which the genes came to be correlated with the traits? Given that they *are* correlated, what difference would it make how they got that way? That's a matter for future research in the new field Morgan named "developmental genetics."

8.3 PROBLEMS IN CHARACTERIZING THE EVOLUTIONARY SYNTHESIS

The Evolutionary Synthesis, as a scientific movement and research tradition, is a large and complex topic of historical and philosophical discussion. Its origins, its status as a scientific theory (or meta-theory, or research orientation, etc.), and its consistency or inconsistency with a large range of alternative evolutionary concepts and opinions have been debated since the publication of *The Evolutionary Synthesis* (Mayr and Provine 1980; Beatty 1986; Burian 1988; Gayon 1989; Dietrich 1995; Reif et al. 2000). The book you are now reading was inspired by one aspect of those debates: the assertion that developmental biology is absent from the Synthesis, and that it ought to be present (Hamburger 1980 and many other writings). Nevertheless, the Synthesis as a scientific and historical entity remains elusive.

An indication of the problematic status of the Evolutionary Synthesis can be seen in Betty Smocovitis's *Unifying Biology* (Smocovitis 1996). This book has a very peculiar form. Smocovitis had written a long and engaging paper on the history of the Synthesis (Smocovitis 1992), and then she began a self-conscious examination of the historiographic methods by which such a history could be written. The book is the result of that examination. It contains the earlier paper as Section 2, described as "The Narrative." Sections 1 and 3 discuss aspects of historiographic and cultural theory involved in doing the history of science. These sections refer to "The Narrative" almost as if it were another book entirely, written by another author. Section 1 includes a chapter title that refers to "Rethinking" the Synthesis, and Section 3 a chapter on "Reproblematizing" it. When I first read *Unifying Biology* I didn't see the point much of the material in Sections 1 and 3. Almost everything of direct historical interest appeared in Section 2. However, after trying to reconstruct the Synthesis myself, I have gained a great deal of respect for Smocovitis's concerns. One important point she makes is that a scientific tradition forms itself, in part, by an active interpretation of its own history. Thus understanding the Evolutionary Synthesis itself requires us to observe the attempts of Synthesis participants to understand their own history.

To acknowledge these difficulties (and prior to a serious attempt to characterize the movement), let me propose a range of possible accounts of the Evolutionary Synthesis.

- The Evolutionary Synthesis originated with the 1937 publication of Dobzhansky's *Genetics and the Origin of Species*, and continued with canonical writings by Stebbins, Simpson, Mayr, Huxley, and Resch. It culminated in an international conference in Princeton in 1947 (reported in Jepsen, Mayr, and Simpson 1949) at which the participants were amazed to find almost complete agreement on basic principles.
- The Evolutionary Synthesis originated with the 1959 Darwin centennial celebrations, at which the agreements in evolutionary theory that had been achieved from 1937 through 1950 were celebrated and (importantly) aligned with Charles Darwin as their intellectual ancestor.

- The Evolutionary Synthesis, as an entity extended in time, was begun in the 1930s and 1940s by the "architects," given more definite form by the 1959 centennial celebrations and the conferences of 1974 that resulted in *The Evolutionary Synthesis* (Mayr and Provine 1980), and strongly shaped by the sometimes-virulent attacks on the Evolutionary Synthesis that began in the 1970s and continued thereafter.[42]

8.4 THE EVOLUTIONARY SYNTHESIS CHARACTERIZED

Despite these historiographic worries, I must give at least a sketch of the nature of the Synthesis in order to examine the tensions that arose with developmental biology. The earliest stages of the Synthesis are least controversial. They involve the development of population genetics and its demonstrated consistency with the MCTH. In the 1920s and 1930s, this overcame what had been a major conflict between Darwinians and Mendelians. The early Mendelians had been mutationists, advocates of discontinuous evolution who believed that selection for continuously varying characters would be impotent to produce long-term evolutionary change. The Darwinians had rejected Mendelism because of its perceived saltationism. Eventually the recognition of a large number of genes of small phenotypic effect, combined with the conceptual distinction between the (discrete) genotype and the (possibly continuous) phenotype, allowed population genetics to be both Mendelian and Darwinian. Population genetics was eventually shown to be consistent with the results of a number of studies of populations in the wild. The MCTH, population genetics, and field studies of variation formed the early core of the Synthesis, and other biological specialties began to reinterpret their theories and results in ways that were consistent with this core.

The theoretical core of the Synthesis was the formal description of populations that was enabled by population genetics. Populations of sexually reproducing organisms were eventually seen as gene pools, the makeup of which changed through time as the result of the changing values of a specific set of parameters. The frequencies of genes in a population vary as a consequence of mutation rate, migration, selection, and drift. In a population in which the values of all of these are zero, there is no genetic change. This is called the Hardy–Weinberg equilibrium. Natural selection is merely one of

---

[42] In Smocovitis's style of reflexive critical awareness, I am tempted to include this very book as an example of attacks on the Evolutionary Synthesis.

a number of parameters, the values of which determine the dynamics of the genetic makeup of populations.

By 1959 the Evolutionary Synthesis was seen as a vindication of Darwin's theory of natural selection. Some aspects of the theory justify this interpretation. However, Gayon shows that natural selection has undergone a massive reconceptualization. During the nineteenth century, heredity and adaptation had been conceived as two great opposing forces. The question was whether the adaptive force (natural selection) had the power to overcome the conservative force (heredity). Population genetics involved the reconceptualization of natural selection. It is no longer a force, or a principle, or even a probability, but merely "a parameter that interacts with a number of others within a homogeneous theoretical field open to many other evolutionary scenarios" (Gayon 1998: 320). The evolutionary force previous thought to work in opposition to selection, namely heredity, has disappeared from view! Heredity (in the form of the MCTH, or in populations the Hardy–Weinberg equilibrium) is merely a background assumption of the entire formal system. Heredity is genes, and the genetic makeup of a population will change only if the parameters change it. This formal characterization of population genetics shows that the commitment to adaptationism must have arisen from somewhere other than population genetics alone. The formalism itself is consistent with scenarios in which migration, mutation, or drift dominate the history of a gene pool. Nevertheless, selection has at last been shown to be *possible*, given the facts of heredity. Prior to population genetics, even this possibility was undetermined.

Population genetics alone could not determine the values of the parameters, and so adaptationism was not built into the framework of the Synthesis itself. The importance of adaptation was a matter of empirical argumentation, and strong arguments about the relative importance of drift or selection continued throughout the century. These are reasonably regarded as internal arguments, with no real possibility of challenging the fundamental assumptions at the core of the Synthesis. However, many other evolutionary concepts, previously popular, were explicitly forbidden by the core Synthesis framework. Some years after coediting *The Evolutionary Synthesis*, William Provine began to emphasize this aspect of the Synthesis:

> The evolutionary synthesis was not so much a synthesis as it was a vast cutdown of variables considered important in the evolutionary process.... What was new in this conception of evolution was not the individual variables, most of which had been long recognized, but the idea that evolution depended on so few of them.... This I will now call the "evolutionary constriction," which

seems to me to be a more accurate description of what actually happened to evolutionary biology. (Provine 1988: 61)

Provine claims that the theoretical "synthesis" of the many biological fields beyond genetics and population genetics was limited to the study of how various factors (studied by specialists) could change gene frequencies. Other than that, he believes that accommodation was achieved not by theoretical unification but by consistency proofs and the removal of barriers between disciplines. This barrier removal involved rejecting evolutionary factors that could not be expressed as population genetic processes. Lamarckian, purposive, orthogenetic, and saltational theories were all rejected as inconsistent with the basic populational mechanisms.

Reif and his colleagues have produced a very useful sketch of the resulting Synthesis commitments (Reif et al. 2000). Their report is pragmatic, objective, and succinct, and it takes account of much of the earlier historical commentary. The paper criticizes the historiography of the Synthesis, but on grounds unrelated to the present discussion.[43] Reif et al. analyze the Synthesis as consisting of five central conceptual components, a set of implications drawn from those components, and a list of concepts that are categorically rejected by Synthesis commitments (Reif et al. 2000: 58 ff.). The five components are these:

- mutations (random with respect to adaptation);
- selection as the primary directional force (and largely restricted to the individual level);
- recombination in sexually reproducing populations;
- isolation (various mechanisms preventing gene flow); and
- drift (the importance of which depends on effective population size).

The implications that "follow automatically from this basic structure" are equally important for understanding the Synthesis:

- Speciation is predominantly allopatric or parapatric (i.e., it requires some isolation among subpopulations).
- Evolution is gradual but can have a wide range of velocities.
- "Developmental, historical and constructional constraints limit the opportunism of evolutionism to a certain degree, but do not lead to non-adaptive evolution." (I quote this factor because it concerns development.)

---

[43] The paper argues that German contributions to the development of the Synthesis have been underappreciated.

The list of concepts definitely rejected by the Synthesis is divided into genetic and macroevolutionary aspects:

- Forbidden genetic factors are macromutations (see, e.g., Goldschmidt 1940) and Lamarckian inheritance.
- Forbidden macroevolutionary factors are a wide range of theories, including progressive and teleological, orthogenetic, "racial senescence," and saltational evolution, and "'Baupläne' or types as actors in evolution" (again quoted because of special relevance).

Reif et al. summarize earlier historiographical discussion. First they acknowledge that the population geneticists already knew the five central factors before 1930. Nevertheless, the "Synthesis proper" includes the work of the architects (especially Dobzhansky) who provided evidence of the actual values in nature of the various parameters. The summary is concluded with two points. First, as shown in Provine's aforementioned discussion, evolutionary factors other than the five were excluded. Second, they report the central (*very* central) Synthesis principle that macroevolution is merely an extrapolation of microevolution. This means that no other factors than populational ones are necessary to account for macroevolutionary patterns.[44] The extrapolation issue will be revisited.

8.5 BY-PRODUCTS OF THE CORE OF SYNTHESIS THOUGHT

I would like to call attention to several other aspects of evolutionary thought that were modified by the Synthesis. Although these are usually not recognized as defining characteristics of the Synthesis itself, they are important to its structure. Among them are systematics, the significance of phylogeny, and the concept of an evolutionary mechanism or cause.

8.5.1 Systematics

The practice called "the new systematics" (Huxley 1940) emphasized categories below the species level, their interactions, and their responses to

---

[44] The actual statement is that "it was demonstrated by extrapolation that the factors acted in macroevolution in the same way as in microevolution" (Reif et al. 2000: 60). This is a highly problematic expression of the issue. Synthesis authors *asserted* that macroevolution was a mere extrapolation of microevolution. The evidence for the assertion was merely that gathered against the relevance of various nonpopulational factors. They did not use a particular inference form called "extrapolation" to establish the adequacy of population genetic mechanisms to explain macroevolution.

selection. Systematics in the nineteenth century had always been concerned with higher taxa and the relations among them. However, if macroevolution was merely an extrapolation of microevolution, higher taxa had no apparent theoretical interest. Ernst Mayr often insists on the importance of systematics to the formation of the Synthesis. When he does so, he means "the new systematics" of variation, populations, and species. He is not referring to the nineteenth-century project of constructing the Natural System. Higher taxa regained their importance only after cladist methods placed their determination on sounder footing in the 1970s and thereafter.[45]

8.5.2   Phylogeny

As with higher taxa, phylogeny had very limited importance under Synthesis theoretical assumptions. Dobzhansky said that nineteenth-century theorists were interested in phylogeny only because the fact of evolution needed additional proof in those days. He apparently could think of no other reason to study it. Bowler has similar views: phylogenetic history was *Life's Splendid Drama*, an interesting narrative that served no important theoretical purpose.

8.5.3   Mechanisms

It is common in evolutionary discussions to distinguish between pattern and process, between the evident diversity of life (pattern) and the causes that made it that way (process). When Synthesis theorists speak of "evolutionary mechanisms," they mean population genetic processes, usually involving selection (but possibly drift or other parameters). Dobzhansky reports that Darwin was one of the very few nineteenth-century thinkers who were interested in "the mechanisms of evolution, the causal rather than the historical problem" (Dobzhansky 1937: 8). This is not the only way to think of mechanisms, just as history is not the only way to think of phylogeny. For Haeckel and Gegenbaur the biogenetic law presupposed a mechanism of phylogenetic change, and the construction of phylogenies was involved in the causal explanation of organic form. The restriction of the term *mechanism* to populational processes is a theory-laden convention. Other mechanisms are associated with other theories, as we shall see.

---

[45] Cladistic systematists are now back in the business of "identifying, naming, and inferring phylogenetic relationships among taxa." Some are resentful of the period of new systematics. It had "redefined the fundamental problem of systematics from discovering the hierarchy of nature to 'detecting evolution at work'" (Brower 2000: 12).

The core of Synthesis theory, as described by Reif et al., includes only a few possible points of interaction with developmental biology. They are (1) the possibility of development constraints, (2) the rejection of bauplans or types, and (3) the rejection of autonomous macroevolutionary mechanisms. Only developmental constraints are listed as a *positive* intersection of developmental and evolutionary factors (i.e., one that is acknowledged by the Synthesis itself). Reif et al. did not intend to list every conceivable factor relevant to Synthesis theory, of course; the by-products I have listed of systematics, phylogeny, and mechanism are additional factors. However, an inspection of this framework of the Synthesis as a theory shows that there is little room for developmental input. Especially if the knowledge of development is expressed nongenetically, it would appear that there is no obvious logical location for development to integrate with Synthesis theory. In Chapter 9 I discuss the reactions of structuralists to this situation.

9

Structuralist Reactions to the Synthesis

9.1 EXPERIMENTAL EMBRYOLOGY AND THE SYNTHESIS

In this chapter I examine the early interactions between embryology and the Evolutionary Synthesis. Section 9.2 sketches some of the accomplishments of experimental embryology. Section 9.3 covers criticisms that some embryologists directed against the Mendelian–chromosomal theory of heredity, and through that the Synthesis. Section 9.4 proceeds to the potentially Synthesis-friendly work by embryologists and other developmental theorists, and the Synthesis reactions to this work. Section 9.5 covers the position of development within the Synthesis up to about 1959.

Scott Gilbert is a developmental biologist, an evo–devo practitioner, and a trained historian of science. Many of his historical writings argue for the relevance of the tradition of experimental embryology to modern biology. Gilbert claims that, during the mid-twentieth century, embryology began to be unfairly depicted as old-fashioned and metaphysically flawed. He traces the beginnings of this disparagement of embryology to two reformed embryologists, William Bateson and T. H. Morgan (Gilbert 1998). The grounds for Morgan's sudden disapproval of his own former field are complex. They surely involved his hopes for the new genetic paradigm, and probably also the institutional and financial support that genetics came to receive in the United States (Allen 1985). I suspect they also involve what I have called Morgan's quasi-positivism. Morgan expresses this influence in a paper on the rise of genetics. In this paper, he separates science from philosophy and metaphysics, which are to be discarded "not because they are wrong, but because they are useless." Morgan aligns pregenetic experimental embryology with (the useless practice of) philosophy: "philosophical platitudes were invoked rather than experimentally determined factors. Then, too, experimental embryology ran for a while after false gods that landed it finally in a maze of metaphysical subtleties" (Morgan 1932: 261, 285).

Morgan was an originator of the historiographic tradition that genetics was the successor (not a partner) to embryology, and that the traditional methods of embryology could contribute nothing to advances in genetics. In 1926 Morgan announced that experimental embryology, in its traditional form, was virtually defunct:

> The study of the fundamental problems of embryology by experimental methods had almost come to a standstill until two new methods of procedure appeared above the horizon – one the direct application of physico-chemical methods to the developing organism; the other, the application of genetics to problems of development. (Morgan 1926b: 510–511)

The new techniques were only "above the horizon." Neither was an active research program. "Morgan was consciously assigning embryology an agenda that was not the agenda that characterized the field" (Gilbert 1998: 174). The statement must be seen as Morgan's vision of the future, a future that will not include the traditional methods of experimental embryology. The irony is that the Golden Age of experimental embryology was in full swing as Morgan was announcing its demise (Oppenheimer 1966; Gilbert 1998: 175). Let us consider the program.

9.2   THE PROGRAM OF EXPERIMENTAL EMBRYOLOGY

Experimental embryology (Entwicklungsmechanik) was the program begun by Wilhelm Roux, though Wilhelm His had advocated a similar program a decade earlier. As we have seen, Roux was an embryological mosaic theorist, and so a preformationist in a restricted sense. Nevertheless, much of the field's progress was epigenetic in nature and focused on regulative rather than mosaic processes. We begin our discussion with the theoretical contrast between Roux and Hans Driesch, his epigenetic counterpart, and proceed through to the beginnings of developmental genetics.

Roux claimed that differentiation in the embryo resulted from the unequal division of nuclear genetic determinates during cell division. In 1888 he reported the results of experiments that demonstrated mosaicism in early frog embryos. Roux killed one of the cells of a two-cell embryo with a hot needle. The other cell continued dividing, and developed up to the neurula stage (at which the neural fold is apparent), but remained as only half of an embryo divided down the body axis. Apparently the surviving half-embryo had only a half-set of hereditary determinates. In 1892 Driesch attempted to extend the experiment by using sea urchin embryos. He found that shaking could disassemble early embryos. To Driesch's surprise, the separated cells of four-cell embryos each developed into a fully differentiated (though smaller) larva. At least for sea urchins, Driesch had proven the totipotency of early individual cells: Each had the potential to differentiate into any part of the body.[46] One cell of an intact four-cell embryo would ordinarily differentiate into about a quarter of the embryo's body. When this ordinary differentiation happens, it is caused not only by the hereditary endowment within that cell but also by the cell's environment – its position in the whole embryo. If the same cell *had been* detached (and thus found itself in a different environment), it *would have* differentiated differently, into an entire larva. This implies a reciprocal causal interaction among the cells as they develop. The cells influence each other's fates as differentiation proceeds.

Driesch demonstrated causal interactions between parts of the embryo. More specific and localized demonstration of this kind of effect came from Hans Spemann's study of the lenses in frogs' eyes. Lenses develop out of the ectoderm that covers the head. They appear in just the right location to fit onto the optic cup, which is formed from underlying neural tissue. In one of a series of experiments that deformed early embryos, Spemann produced a cyclops embryo. He was surprised to find that the single lens fit perfectly onto the single eye. Because the cups and lenses of eyes are built of tissues that are not even in contact during early development, how did they manage come together to form an integrated eyeball, even when the eyeball is in an unusual position on the head? In results first published in 1901, Spemann showed that the optic cup itself stimulates the ectoderm that eventually covers it to differentiate into a lens (Saha 1991). He did this by destroying the precursor of the optic cup on one side of an embryo. The lens failed to form on that side, but it formed normally on the side that had an intact optic cup. Spemann later transplanted the precursors of optic cups to various locations under the ectoderm. Lenses were induced to form in ectoderm that would ordinarily come to be located on the head. However, the further his transplants were located from the usual site of the eye, the less likely that its covering ectoderm

---

[46] Roux's failure to produce this result is usually blamed on the fact that he was unable to separate the killed cell from the live remaining cell of the frog embryo. Although Roux's inference may have been overeager, the actual developmental differences between the organisms chosen by different experimenters has been extremely influential in the theoretical points that have been defended.

would form a lens. Spemann referred to "circles of diffusion" of lens-forming potential, radiating out from the ordinary lens site. Areas of developmental influence like this eventually were termed *morphogenetic fields*.

The lens cup and the ectoderm that responds to it compose a system, not merely a cause and a mechanical effect. C. H. Waddington distinguished between induction (Waddington preferred the term *evocation*) and competence: The results of an induction depend both on the nature of the inducer (here the transplanted optic cup) and on the competence of the induced tissue to respond (Waddington 1940). The competence of a tissue itself changes through time as a result of earlier inductions. It is now known that head ectoderm gets its lens-forming competence through the history of its tissue movements during ontogeny. During the gastrula stage, it is in contact with underlying endoderm; it later comes in contact with cardiac mesoderm (material that will develop into the heart); and finally with the optic cup. The first two contacts induce the competence in the head ectoderm to eventually respond to the optic cup (Gilbert 2003b: 146; Jacobsen 1966).

His study of lens induction prepared Spemann for the achievement that earned him the 1935 Nobel Prize, the discovery of the embryonic area within the amphibian gastrula called *the organizer*. Spemann wanted to identify the places and times in development where the fates of cells became determined. He began a series of experiments in which sections of ectoderm were transplanted from place to place on the surfaces of gastrulas. Transplants in the early gastrula stages would take on the character of their new location, whereas later transplants would retain the character of the location they were taken from. Determination had taken place between those times. What caused it? Together with his student Hilde Mangold, Spemann began to make xenoplastic (cross-species) transplants between embryos of different coloring. This made it possible to identify the host and donor tissues even after the transplant had been incorporated into the host. Eventually a portion of the dorsal lip of the blastopore (the opening to the inside of the gastrula) of one embryo was transplanted onto another embryo (which of course already had its own blastopore). The transplanted lip first caused the creation of a second blastopore in the host. It then induced the formation of an entire second neural tube and virtually an entire second embryo. The doubly neurulated host grew into what looked like two conjoined embryos, belly to belly (Gilbert 2003b: 320), as shown in Figure 7.

The difference in pigment made it possible to determine the source of the tissues in the secondary embryo. Only a small portion came from the donor. That small bit of donor tissue had induced the formation of almost an entire new organism within the tissues of the host embryo. In a paper published in

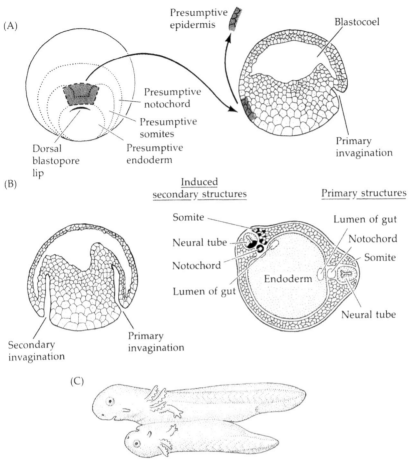

Fig. 7. Spemann and Mangold's 1924 organizer experiment: The dorsal blastopore lip from an early gastrula (A) is transplanted into another early gastrula in the region that normally becomes ventral epidermis. (B) Tissue invaginates and forms a second neural axis. Both donor and host tissues are seen in this new neural tube, notochord, and somites. (C) Eventually a second embryo forms that is joined to the host. As crucial as it is to development, the organizer was not genetically characterized until the early 1990s (Smith and Harland 1992). Transmission genetics could not touch it. From Gilbert 1988: 348. Reproduced by permission of Sinauer Associates.

1924 (after Mangold's accidental death), the donated tissue was named *the organizer*.

Not all embryological processes derived from regulative interactions between cells. Organisms such as tunicates and nematodes are highly mosaic in their development, with the cells derived from certain portions of the

173

egg destined to form certain body parts. Even the embryos of sea urchins, Driesch's experimental animal, show mosaic influences that Driesch's experiments didn't reveal. The single-cell urchin already has an axis, with what is called the animal pole at one end and the vegetal pole at the other. In an elegant series of experiments around 1930, Sven Hörstadius demonstrated the complex influences within the developing embryo of these preformed areas. Hörstadius began by separating eight-cell urchin embryos in half. If he did so meridionally (so that each half was made up of two animal and two vegetal cells), he got small but well-formed pluteus larvae. If he did so equatorially (separating the animal half from the vegetal half), he got one deformed larva and one simple ball of ciliated cells (called a *dauerblastula*). He extended the experiments to the sixty-four-cell stage, at which the embryo is made up of six tiers of cells (three tiers of animal cells, two tiers of vegetal cells, and a tier of micromeres at the very tip of the vegetal pole). Each of these tiers was known to give rise to particular parts of the body under ordinary developmental circumstances. Hörstadius separated the various tiers and recombined them in different combinations. Certain combinations of tiers produced nearly normal morphologies, with the "missing" body parts regulatively produced from cells of the remaining tiers. Other combinations gave rise to dauerblastula and other malformations. The difference between the pluteus morphologies and the malformations was not due to the *particular* tiers that were combined. Instead, the well-formed larvae seemed to develop from combinations of tiers in which both the so-called animal and vegetal characters were equally well represented. A well-formed pluteus could be formed from the most extreme vegetal tier combined with the most extreme animal tier, or with a pair of more medial animal and vegetal tiers. It appeared that the production of a normal morphology required not simple determinates for each part, but rather a pair of opposed gradients, one animal and one vegetal. The animal–vegetal gradient is established in the fertilized cell, and later cells have their developmental potential restricted by their lack of both gradients. Here is another application of the concept of the morphogenetic field; the field itself (whatever cell layers it was composed of) would allow the development of the pluteus larval form. In 1892 Driesch was fortunate to have chosen four-cell embryos for his experiments: All four cells of an urchin embryo have animal and vegetal poles. The next cellular division is equatorial, which separates the animal from the vegetal poles of each of the four cells. Driesch's experiment that demonstrate regulation would have failed with eight-cell urchin embryos.

The purpose of this section is to convey the flavor of experimental embryology as it was practiced in its golden age. The holism of the era is apparent in the concept of the morphogenetic field (Gilbert et al. 1996). The concept of morphogenetic fields seemed directly opposed to the atomism of MCTH genetics. Even with the recognition by the geneticists that genes and characters have a many–many relationship, that relationship is only *between genes and characters*, with no apparent room for such entities as fields, or even inductive interactions. It is hoped that this brief sketch will allow the reader to appreciate the skepticism that the embryologists would show toward the MCTH and a fortiori for the evolutionary theory that was based on it. How could a hypothesized particle be "the cause" of an eye color, when even the location of the eye itself was the result of complex inductive interactions among the various parts of the developing embryo?[47]

9.3   THE EMBRYOLOGICAL CRITIQUE OF THE SYNTHESIS

Developmentally inclined critics of mainstream evolutionary theory (including me) have sometimes given the impression that a conscious conspiracy had kept development out of the Evolutionary Synthesis. In contrast, Ernst Mayr has stated that embryologists simply weren't interested in participating (Mayr 1991: 8). The historical record shows that Mayr was correct, although the story is complex. Tensions have always existed between Synthesis biologists and those who consider development to be important to evolution. In this section I address a number of developmental reactions to the Synthesis (and to the genetic views associated with the Synthesis) in the period prior to about 1955. I begin with a principle that has again become prominent with the evo-devo movement.

9.3.1   Critique 1: The Causal Completeness Principle

The notion I call the Causal Completeness Principle was so commonly accepted prior to the Synthesis that it was rarely even enunciated. It was a presupposition of almost every nineteenth-century evolutionary theory. We have seen its precursors in the developmentalist doctrine of evolutionary change (Chapter 4, Section 4.4 and 4.9) and the Generative Rule of phylogenetic

---

[47] This conflict between development and genetics was not short lived. In 1997, Dick Burian wrote about the "apparent rapprochement, now underway" between development and genetics. Here is his summary of the history of the conflict: "[T]he disagreements were based in part on the absolute inability of geneticists to show how genes could account for the *Bauplan* of an organism and on their failure to give any weight to such phenomena as cytoplasmic gradients in the egg, polarities in the egg, and the early embryo, cell death in organogenesis, and so on" (Burian 1997: 243).

reconstruction (Chapter 5, Sections 5.4 and 5.7). It became rhetorically important only when a theory arose that contradicted it. In the nineteenth century, virtually nothing contradicted it, not even the preformationist views of Roux and Weismann – but the Evolutionary Synthesis contradicted it.

> The Causal Completeness Principle: In order to achieve a modification in adult form, evolution must modify the embryological processes responsible for that form. Therefore an understanding of evolution requires an understanding of development. (After Horder 1989: 340)

This principle or a variant has been attributed to Richard Goldschmidt and C. H. Waddington (Gilbert et al. 1996), to Walter Garstang (Raff 1996), to Gavin De Beer, and even to Jackson St. George Mivart (Johnston and Gottlieb 1990). I was unable to locate Mivart's use, but I have no doubt that he and most of his contemporaries believed it. As we have seen, heredity *was* development in those days. Development was entwined with evolution, because evolution necessarily involved heredity and heredity necessarily involved development.

In 1915 Morgan and his coauthors innovatively separated heredity from development, and they insisted that heredity nevertheless *causes* traits. The Evolutionary Synthesis parlayed this new kind of causation into a causal explanation of evolution that bypassed development. Those who considered development important for evolution now had a reason to say so. Frank Lillie was one of the embryologists who were skeptical about genetics even prior to the Synthesis. He observed that the methods used by geneticists "have no place among their categories for the ontogenetic process and *a fortiori* for the phylogenetic" (Lillie 1927: 368). In other words, the fact that genetics does not deal with development implies (to Lillie) that it cannot deal with evolution. A decade later, Dobzhansky would clearly state the converse, in a principle that would become the methodological basis of the early Evolutionary Synthesis:

> Since evolution is a change in the genetic composition of populations, the mechanisms of evolution constitute problems of population genetics. (Dobzhansky 1937: 11)

The Causal Completeness Principle is better understood as a statement of theoretical commitment rather than an actual argument. Horder's and Dobzhansky's aforementioned statements contrast nicely. Each declares both an explanans of evolutionary theory ("modification of adult form" vs. "change in the genetic composition of populations") and a sketch of the proper explananda ("how embryological processes can be modified" vs. "problems of population genetics"). The contrast is not simply in how evolution should be explained, but also what *about* evolution is worth understanding. Horder's favored explanans (form) is traceable to the early morphologists in the nineteenth century. Dobzhansky's explanans (change – now defined as change in the genetics of populations) goes back to Darwin.

We now turn to developmental critiques of genetics and of the Synthesis that were more openly in play between the 1915 announcement of the MCTH and the mid-1950s. It will be obvious that the advocates of development accept the Causal Completeness Principle even if they don't enunciate it.

9.3.2 Critique 2: The Developmental Paradox

The Developmental Paradox expresses (what the embryologists saw as) the impossibility in principle for MCTH genetics to explain ontogenetic development. It is best understood by comparison with the judgments of early-twentieth-century embryologists toward the Weismann–Roux theory of mosaic development. On Weismann's view, the progressive differentiation of somatic cells in the body was a product of the unequal distribution of genetic determinates in successive cell divisions during ontogeny. The theory had two aspects: the explanation of the fact of differentiation itself (the production of an embryo that gradually became more and more differentiated through its ontogeny), and the explanation of the actual properties of the differentiated body parts (the boniness of bone, the contractility of muscle, etc.).[48] Weismann's theory successfully accounted for the fact of differentiation (assuming unequal genetic distribution). However, the qualities of the differentiated cells and body parts were left unexplained except in a circular manner: The bony nature of bone cells is caused by whatever genetic determinates are distributed to bone cells; the same is true for muscle, nerve, and so on. Most early-twentieth-century embryologists rejected the Weismann–Roux theory as preformationist. They did so *not* because it failed to explain the fact of differentiation – that was explained. They labeled it preformationist because it failed to account for (or accounted for circularly) the properties of the differentiated parts.

By 1910 it was generally accepted that the Weismann–Roux theory had an additional problem. Not only was it methodologically flawed by its preformationism, it was empirically refuted on cytological grounds. Empirical evidence indicated that cellular division was always equal, and all genetic material was passed to each daughter cell after a division, so the empirical failure of its explanation of differentiation was added to its methodological failure

---

[48] Lillie describes this distinction as "differentiation in its two aspects of embryonic segregation of potencies and of realization of potencies" (Lillie 1927: 362).

with respect to the properties. In 1910, in his last anti-Mendelian publication, T. H. Morgan called attention to this failure:

> We find that the chromosomes in the different tissues are identical as far as our methods reach. Observation gives a positive denial to the Roux–Wiesmann [sic] assumption.... I myself have found the same disinclination to reduce the problem of development to the action of specific particles in the chromosomes.... [I]t is unsafe and unwise to reduce the problem of heredity and development to a single element in the cell; when we have every evidence that in plasm is the real seat of the changes going on at this time, while the chromosomes remain apparently constant throughout the process. (Morgan 1910)

How could development, which is essentially a process of change in cells, be explained by particles that remain unchanged in every cell?

This was Morgan in 1910. We have already seen his 1915 espousal of the MCTH. Let us now meditate on how badly the MCTH violated the intuitions Morgan himself expressed in 1910. The MCTH acknowledges that the genetic material is the same in all body parts. It nevertheless asserts that a single genetic element (p), an element that exists in every cell of the body, is nevertheless the cause of the pink color *that exists only in the cells of the eye.* Such a move was "unsafe and unwise" in 1910, but it was "the sense in which science always uses this expression [cause]" in 1915. (I remain fascinated with the methodological shift that allowed Morgan to make this conceptual leap.)

The Weismann–Roux theory had been judged preformationist because of its failure to explain the properties of differentiated parts, but not with respect to its explanation of differentiation itself. The MCTH retains Weismann–Roux's preformationism regarding properties, and it adds to it a new preformationism regarding differentiation itself. It is exactly twice as preformationist as Weismann's theory! Morgan fully acknowledges that the MCTH does not explain development (differentiation), of course. From the embryological point of view, though, that admission doesn't help. With genes directly causing traits, the MCTH appears to leave no room for development even to take place:

> With reference to the processes of embryonic segregation, genetics is to a certain extent the victim of its own rigor. It is apparently not only sound, but apparently almost universally accepted genetic doctrine to-day that each cell receives the entire complex of genes. It would, therefore, appear to be self-contradictory to attempt to explain embryonic segregation by behavior of the genes which are *ex hyp.* the same in every cell.... Those who desire to make genetics the basis of physiology of development will have to explain how an unchanging complex

can direct the course of an ordered developmental stream. (Lillie 1927: 365, 367)

The MCTH leaves us with the Developmental Paradox, also known as "the paradox of nuclear equivalence during cellular differentiation" (Sapp 1987: 17) and "Lillie's Paradox" (Burian 2005). Genes determine the nature of each body part, and body parts differ from each other, but each body part holds the same genes. Lillie's final sentence in the aforementioned quote shows why the problem is most striking to embryologists: Like all epigenesists since von Baer, their job is to explain the changes from homogeneity to heterogeneity. The geneticists have given them the genome with which to work, but the genome *doesn't change* during development – it shows no increase in heterogeneity. How can unchanging genes explain changes during development, or (to put it another way) how can identical genes be the causes of differentiated body parts? The paradox might not have seemed so troublesome if the embryologists had not themselves already had some success at experimentally identifying the nongenetic determinates of differentiation. Cell fates could be changed by transplantation or by introduction of foreign materials, and bodily axes were experimentally demonstrated to be preestablished in the egg. The entire body of work sketched in Section 9.2 militates against the notion that hereditary determinates can be conceived as the "causes" of adult traits in the absence of the causal complex of embryological development.

The paradox was commonly acknowledged in the early days of genetics, and no attempt was made to hide it. Morgan's 1910 statement cites the paradox as grounds to reject particulate theories. He cites it again in his 1933 Nobel Prize Lecture, but by then he treated it as one of the open questions of genetics: "Every cell comes to contain the same kind of genes. Why then is it that some cells become muscle cells, some nerve cells, and others remain reproductive cells?" (Morgan quoted in Sandler and Sandler 1985: 369.) Almost thirty years later, in a landmark paper on the operon model of gene regulation, Jacques Monod and François Jacob described their achievement as resolving the paradox. They were working on bacteria, which had previously been regarded as irrelevant to the development of multicellular organisms. They argued that enzymatic adaptation in individual bacteria was analogous to differentiation among the cells of metazoa. They proposed an updated version of Roux–Weismann mosaicism as their stalking horse: "That differentiation [within metazoa] involves induced . . . alterations of the genetic information of somatic cells has often been proposed as the only possible interpretation of the 'paradox'" (Jacob and Monod 1961: 400). They offered their genetic explanation of the adaptation of bacteria as a solution of the Developmental Paradox within metazoa: "[B]iochemical differentiation... of cells carrying an identical genome, does not constitute a 'paradox,' as it appeared to do for many years, to both embryologists and geneticists" (Jacob and Monod 1961: 397). I know from classroom experience that the Developmental Paradox retains its ability to perplex today. Educated nonbiologists, who have assimilated the popular metaphors of the human genome as a blueprint, a code, or a Book of Life, are taken aback when asked to explain how cells with the same genome have become differentiated in the body.

Geneticists and embryologists differed on what the paradox implied about future research. Geneticists were convinced that heredity was nothing but transmission genetics, and the paradox was merely a puzzle to be solved in the related domain of developmental genetics. Even though a relatively small number of genes had been identified, the physiological study of genes (which had been identified transmissionally – there was no other way to identify them) would eventually resolve the paradox. Embryologists simply doubted that transmission genetics was all there was to heredity. The notion that genes somehow "caused" adult traits – traits that crucially depend on embryonic differentiation – while admitting that no one could explain how genes could produce *differentiation itself*, was more than many embryologists could accept. They proposed alternative systems of heredity. They distinguished between the *superficial* or *individual* characters (caused by Mendelian factors) and the *fundamental* or *generic* characters (controlled by the alternative heredity system; see Sapp 1987: 16 ff.). They had surprisingly reasonable grounds for doing so.

9.3.3   Critique 3: Fundamental Versus Superficial Characters

For many early students of heredity, one of the most convincing features of the MCTH came from linkage groups and their correlation with chromosome number. Morgan's school determined that mutations did not segregate fully independently in a Mendelian fashion. Traits seemed to be linked with others, and inherited together with various probabilities. The MCTH proposed that genes are positioned on pairs of chromosomes, which sometimes "crossed over." The crossing over would explain linkage patterns. Morgan pointed out that, in *Drosophila melanogaster*, there were four such linkage groups and four chromosomes. Historian Stephen Brush studied the features that influenced the acceptance of the MCTH in the United States and Britain, and he identified the six factors that were most important in the theory's acceptance. The correspondence between chromosome number and linkage groups was the single factor mentioned by the largest number of converts. Brush is surprised to recognize how quickly this occurred:

> Since this [chromosome] number varies for other species than *Drosophila*, one might have thought that the linkage group theorem should have been verified for more than just one species (or the several species of *Drosophila*) but that did not seem to be crucial to accepted the universal validity of the MCTH. (Brush 2002: 517).

I report Brush's results only to illustrate that broad theoretical considerations can make the demand for inductive confirmation fade quickly into the background. The MCTH was so theoretically powerful (to its converts) that a single correlation between chromosomes and linkage groups was enough to support the "universal validity" of the theory. Embryologists shared a different set of presuppositions. Even if the linkage–chromosome projection was extensively confirmed with other species (as of course it was), they remained skeptical about the universality of the theory. They were concerned not whether Mendelism applied to *all species*, but rather whether Mendelism constituted *all of heredity*, even within a single species. There were reasons to doubt it.

The distinction between fundamental and superficial characters was usually aligned with the contrast between cell nucleus and cytoplasm. Even though it could be demonstrated that nuclear material conveyed determinates for some characters, embryologists were very reluctant to acknowledge that this was all of heredity. They had several grounds for this.

The first was the Developmental Paradox itself. Embryologists felt that the solution to the paradox must lie in an understanding of the causal role of the cytoplasm in the overall development of the organism (as Morgan said in 1910, "we have every evidence that in plasm is the real seat of the changes"). Whatever cytoplasmic mechanism is involved in the solution to the paradox should be considered an additional form of heredity. After all, that mechanism (and not only the nuclear material with which it interacts) must be inherited too.

Second, the characters studied by early geneticists were known to be a small and very biased set of the characters possessed by any organism. This was not only because the field was young. It was also because the methods of Mendelian analysis had systematic biases that excluded the very characters that were most important to embryologists. Until the late 1930s, almost all genetic research was based on breeding experiments. The operational identification of a gene was accomplished by crossing two individuals that varied with respect to the trait in question. In order to discover the genetic basis of a trait, the geneticist must obtain two individual organisms that (a) varied in their possession of the trait and (b) could be interbred. If either of these conditions could not be met with respect to a trait, its genetic basis was undiscoverable. Given the fact of interspecies sterility, this creates two important biases in the sample of characters available for genetic study. I call these the *Mendelian blind spots*:

1. Mendelian breeding experiments cannot study characters that are fixed within a species, because no variants exist to cross.[49]
2. Mendelian breeding experiments cannot study characters that vary only between species (or between higher taxa) because of the sterility of such crosses.

The blind spots prohibited geneticists from operationally identifying genes for traits fixed within species, or genes for traits that varied only between taxa. Geneticists were forced by their own methods to experiment only on characters that vary within a species. It was quite natural for embryologists to consider these to be superficial characters. The label *superficial* was not mere rhetoric. The Mendelian methods ruled out the study of every embryological character that had ever been involved in the Unity of Type. Those were important (indeed *fundamental*) characters to embryologists. This is the context for E. E. Just's much-quoted 1937 jibe that he was interested more in the fly's back than the bristles on its back, and more in its eye than its eye color (see Chapter 4, Section 4.9). Geneticists had chosen to study bristle number and eye color not because of their intrinsic importance as characters. They chose to study those characters *only because they could*, because intraspecies variability existed. All of the geneticists' examples came from such characters.

The Mendelian blind spots indicated (to embryologists) that it was impossible to test the universality of the MCTH by the use of genetic experiments alone. If there were a second system of heredity, geneticists would never be able to find it. Oscar Hertwig had protested the preformationism of Weismann in 1894. His daughter Paula Hertwig protested the preformationism of the MCTH forty years later. She contrasts the genetic view with the

---

[49] The experimental production of mutations by radiation and other means (mutagenesis) did gradually allow identification of recessive lethal genes that were not otherwise observable in the population. Opinions differed greatly between embryologists and geneticists about whether this method produced a reliable sample of the genes (or other factors) that might influence development.

"dualist" view that proposed both fundamental and superficial (genetic) forms of heredity:[50]

> Can we refute one or the other of these positions by experiments at this time? Not by means of *classical Mendelism*. For in order to recognize or to localize genes, we need two organisms that differ in some particular traits, and even if we can trace certain fundamental properties ... back to Mendelian genes, it appears to me to be completely impossible to remove the force of the dualists' arguments by analysis of hybrids. Johannsen put it like this: "*By analysis of hybrids we have examined only the clothing*, the underlying organization remains unanalyzed. Whether we will ever be able to strip [the species, genera, etc.] of their superficial characteristics in such a way as to reveal the ultimate X of our formulas – a fundamental substance, something quite general and organic, something that, like a homozygote, is not accessible to Mendelian analysis – that remains an unanswerable question." (Hertwig 1934: 428; emphasis in the original)

Johannsen's reference to homozygotes (quoted in Hertwig 1934) refers to the impossibility of obtaining genetic information by crossing homozygotes for an allele. Such crosses give no information about the effect of the allele on the phenotype. Nevertheless, such an allele *does affect* the phenotype. The problem is that Mendelian methods prohibit the discovery of its effect. This illustrates the first Mendelian blind spot, which is the genetic inaccessibility of traits that show no variation within a population. If the gene is fixed in the population – if all members are homozygotes for the gene – it is invisible in all crossing experiments. The arbitrariness and triviality of the characters that *happened to be* amenable to Mendelian treatment made embryologists feel quite justified in hypothesizing a different heredity source, what Hertwig called the "ultimate X," for these hypothetical non-Mendelian, fundamental characters.

Overall, embryologists (1) had never been very interested in characters that vary within a species; (2) had always been interested in characters (such as limbs, or hearts, or cleavage patterns) that were shared across large segments of the taxonomic tree; and (3) were extremely interested in the ontogenies of those characters.

Geneticists offered them (1) genetic analyses of species-variant characters only, (2) with no operational possibility of identifying genes for traits shared by higher taxonomic groups, and (3) a refusal (for the foreseeable future) to account for the ontogenies of the characters for which they identified genes.

---

[50] This translation is by Richard Burian (Burian forthcoming). I thank Freitson Galis for helping me understand this passage.

Given the geneticists' offer, it made good sense to the embryologists to hold out for something better.

A third ground for the fundamental–superficial distinction was the fact that, for many years, the known genes all acted late in development. The first early-acting gene was for chirality (handedness) in snails. Embryologist Klaus Sander points out that this gene has no other effect than handedness, so it is morphologically trivial (Sander 1985: 367). As time passed, more and more early-acting genes were discovered by mutagenesis (e.g., X-ray bombardment). These were typically lethal, but they were traced to events in early development (such as the lack of development of an alimentary canal). Both geneticists and Synthesis evolutionists took this as proof of genetic universality and the illegitimacy of the fundamental–superficial distinction. It did indeed refute those who stated their case only against early gene action, or complained that genes seemed to cause only adult characters. The structuralist's complaint that geneticists (and Synthesis evolutionists) are interested only in adult characters has persisted, and it has persistently been rejected. Taken literally, it is false. Embryonic characters can be analyzed genetically, and they can be analyzed adaptively. Nevertheless, embryologists (and structuralists in general) have a special interest in embryonic traits than is not shared with transmission geneticists and evolutionists – the causal role of embryonic traits in ontogeny. Adult manifestation is not the only problem. Another is this: Characters that were said to be caused by genes are conceived as stationary, so to speak. They are, by Morgan's original definition, uninvolved with causation, taken out of the causal loop. Lillie expresses this grievance:

> [A]t whatever stage of development a character may be selected for examination, and whatever the nature of the character, it must always, so far as the genetic method is concerned, be treated as a finality. It has no past, except the genes postulated as a result of their appearance in previous generations – and no future. The genetic method reveals *alpha*, the gene, and *omega*, the final term. (Lillie 1927: 367)

So even if this omega happens to be an embryonic character, the embryonic *nature* of the character (its causal role in ontogeny) is left out of the analysis. Lillie's critique thus ties the problem of embryonic characters back into the Developmental Paradox. Even if geneticists succeed in identifying genes for embryonic traits, those traits are ripped out of their ontogenetic context and treated as mere characters. Embryonic characters are often causally related *to each other*, as the embryologist sees it – they interact in the course of ontogeny, to produce a three-dimensional organism – or rather a

four-dimensional organism, one that develops through time. So the demonstration of a genetic cause for embryonic characters does not satisfy the embryologist who complains about the emphasis on adult traits. Embryologists'
affection for embryonic characters comes not just from the time of appearance
but also from their causal role in ontogeny.

9.3.4 Cytoplasmic Inheritance Versus Darwinian Extrapolation

Jan Sapp's *Beyond the Gene* documents the long debate regarding cytoplasmic inheritance during the twentieth century (Sapp 1987; see also Thieffry
1996). The early discussions were primarily genetic, but conflicts with the
Evolutionary Synthesis were eventually recognized. Morgan had rejected cytoplasmic inheritance in *Mechanisms of Mendelian Inheritance*, primarily on
the ground that there was no proof of physical continuity among cytoplasmic
materials (Morgan et al. 1915: 135 ff.). In 1919, Morgan claimed that his
researchers had ample empirical evidence against cytoplasmic inheritance:

> Mendelian workers can find no distinction in heredity between characteris-
> tics that might be ... fundamental and those called 'individual' [superficial].
> This failure can scarcely be attributed to a desire to magnify the importance
> of Mendelian heredity, but rather to experience with hereditary characters.
> (Morgan 1919: 126)

This assurance was unacceptable to the embryologists who doubted the universality of the MCTH, of course. The Mendelian workers' "experience with
hereditary characters" was restricted by the Mendelian blind spots already
discussed. Genetic methods could not refute the existence of "fundamental"
characters because the methods couldn't possibly discern them if they existed.

Sapp comments that Morgan's rejection of cytoplasmic "fundamental"
heredity relies on what he calls "the Darwinian view that the nature of hereditary differences between species could be elucidated by studying heredity in
crosses within species" (Sapp 1987: 29). The comment is slightly anachro-
nistic. The general idea that variations within a species contribute to species
change is certain Darwinian to the core. However, the study of heredity by
crossbreeding was a Mendelian idea, not a Darwinian one. Nevertheless, it
was soon to be Darwinized by the Evolutionary Synthesis.

The beginning of the twentieth century had seen a dramatic conflict between followers of Darwin and those of Mendel. The Darwinians were biome-
tricians, and they were committed to hereditary continuity of variation. The
Mendelians were saltationists and mutationists (Provine 1971). This conflict was remedied by the recognition of small Mendelian variations and the

development of mathematical population genetics. After the problem of Mendelian saltationism had been solved, Darwinism and Mendelism seemed a marriage made in heaven. The Mendelian blind spots are irrelevancies for a Darwinian understanding. The operation of natural selection relies on hereditary variation within a breeding population. Although it might be *nice* to study the genetic variations between species and between higher taxa, such studies are (by hypothesis) irrelevant to natural selection. Hereditary factors that are fixed within a breeding population are also irrelevant to natural selection; if it doesn't vary, it can't be selected! What the embryologist sees as a blind spot of Mendelism, the Darwinian sees as an irrelevancy.

There was still room for skepticism of the new Synthesis, of course. The MCTH shares one problem with Darwin himself. Darwin used the analogy of artificial selection to exemplify the actions of heredity under selection. Like artificial selection, Morgan's breeding and mutagenesis experiments were never able to produce new species. However, if *in fact* the processes of microevolution (natural selection of small Mendelian variations within breeding populations) were the only mechanism by which evolution occurs, then it would be perfectly proper to "extrapolate" the results of macroevolution from microevolutionary processes. This is exactly what the Synthesis did. Early statements were cautious.

> [W]e are compelled at the present level of knowledge reluctantly to put a sign of equality between the mechanisms of microevolution and macroevolution, and proceeding on this assumption, to push our investigations as far ahead as this working hypothesis will permit. (Dobzhansky 1937: 12).

However, with the expansion and "hardening" of the Synthesis, confidence in extrapolation increased. In the third edition of Dobzhansky's book, the "working hypothesis" was promoted to a near certainty: "The words 'microevolution' and 'macroevolution' are relative terms, and have only descriptive meaning; they imply no difference in the underlying causal agencies." The consistency arguments from allied biological fields had been coming in. Dobzhansky was able to cite books on paleontology (Simpson), comparative anatomy (Rensch), and embryology (Schmalhausen) that implied that "nothing in the known macroevolutionary phenomena... would require other than the known genetic principles for causal explanation" (Dobzhansky 1951: 17).

Dobzhansky's rejection of cytological inheritance hardened in a similar manner. In 1937 he began a criticism of the fundamental–superficial distinction that began "It has been contended..." and ran for about a page (Dobzhansky 1937: 19). Dobzhansky complained that advocates of the distinction failed to give criteria for the fundamentality or superficiality of characters. He noted that mutagenesis experiments had disclosed lethal recessives that caused an embryo to fail to develop an alimentary canal. He then sarcastically asked whether the lack of an alimentary canal was "fundamental" enough for the advocates of cytoplasmic inheritance. The 1951 version was much abbreviated. It simply reports that certain factors had *"in the past* given rise to the contention that mutations affect only 'superficial' but not 'fundamental' traits" (Dobzhansky 1951: 31; emphasis added).

Genetics clearly dominated the field by this time, and most of the advocacy of cytoplasmic heredity had faded. The issue, however, was still not dead. Boris Ephrussi was a Russian–French embryologist who spent two years in the mid-1930s in Morgan's lab, which had moved in 1928 to the California Institute of Technology in Pasadena. He was one of the earliest workers on physiological genetics, and he was influential both on George Beadle and Jacques Monod, both of whom received Nobel Prizes for their work in genetics (Burian et al. 1991; Amundson 2000). Ephrussi was persistently interested in cytoplasmic inheritance, and his relations with the Morgan group made him very aware of his disagreements with the Synthesis assumptions about heredity. Ephrussi had discovered the cytoplasmic basis of respiration in yeast (surely a "fundamental" character, he thought), and he tentatively identified it with mitochondria, organelles within the cytoplasm (Ephrussi 1951). As late as 1953, Ephrussi defended the fundamental–superficial distinction against Dobzhansky's arguments of 1937. He points out that the respiration discovery was only possible because yeast can actually survive without respiration! The discovery could not have been made by genetic methods, or in an organism in which the loss of respiration would have been lethal. He quotes Dobzhansky's entire 1937 criticism of the superficial–fundamental distinction and responds in detail. As for Dobzhansky's list of early lethal mutations proving the existence of corresponding genes, Ephrussi points out that the same factor (lethality in mutants) might be the cause of difficulty in identifying particular "fundamental" cytoplasmic traits (Ephrussi 1953: 120). Unfortunately, Ephrussi was unaware that Dobzhansky had already pronounced the fundamental–superficial debate to be a thing of the past. Dobzhansky's commitment to extrapolation and his rejection of cytoplasmic inheritance had both hardened by 1951, and cytoplasmic inheritance was dismissed in two sentences.

Genetics was a burgeoning field, and, from the Synthesis standpoint, there was no reason to doubt that genetics comprised all the heredity that mattered to evolution. Embryology was already fading in reputation, in the face of the increasingly molecular studies of development. To embryologists, though, the Synthesis was far from complete. It had two related gaps: the failure to deal with ontogeny, and the inability to genetically study the traits that characterize species and higher taxa. Moreover, these two gaps are conceptually related. It remained possible that some as-yet unknown factor, call it Factor X (following Hertwig), would explain both the species-invariant generation of form within the embryo and the broad non-Mendelian "inheritance" of form among taxa (i.e., Unity of Type). The inability of genetics to explain ontogeny could be seen as the flip side of the invisibility of heredity in phylogenetic comparisons. If Factor X were somehow discovered, it might explain both ontogeny and Unity of Type.

Stephen Jay Gould originally referred to the "hardening of the Synthesis" to describe the increasing adaptationism between the late 1930s and the 1950s (Gould 1980, 1983). Another aspect of this hardening was the increasing opposition to the embryological alternatives such as cytoplasmic inheritance. Viktor Hamburger, a student of Spemann who was working in embryology in the 1930s, was a contributor to the conferences reported in Mayr and Provine (Mayr and Provine 1980). He criticizes the Synthesis evolutionists for "black-boxing" development (Hamburger 1980). As far as I can tell, it was never the conscious intention of Synthesis advocates to oppose the discussion of development, but the perceived relevance of development weakened as the perceived universality of the MCTH increased. The Extrapolation Principle seems to be at the center of the rationale. If microevolutionary population genetic processes in fact extrapolate neatly into macroevolutionary results, then development *really is* irrelevant to evolution. Development is irrelevant to MCTH heredity, MCTH heredity is at the core of population genetics, and population genetics constitutes microevolution. Therefore, if microevolution extrapolates neatly into macroevolution, then the Developmental Paradox and the obscure possibility of nonnuclear inheritance must be irrelevant to the evolutionary process.

From the embryological viewpoint, this reasoning is backward. The Darwinian Extrapolation Principle can only be proven by demonstrating that the MCTH is universal and no fundamental process exists in addition to genetics. However, the universality of the MCTH can only be proven by showing that so-called fundamental heredity is not needed to explain ontogeny. This means that ontogeny must be explained (in a form consistent with the universality of MCTH genetics) *before* the Extrapolation Principle can rationally be accepted. In other words, the black box of development must be opened before the Extrapolation Principle itself can be justified. To use the Extrapolation Principle to reject nongenetic inheritance simply begs the question.

So much for the antagonism between developmental and Evolutionary Synthesis frameworks. Let us turn to some potentially cooperative relations.

9.4 POINTS OF CONTACT AMONG DEVELOPMENTAL AND GENETIC BIOLOGISTS, AND SYNTHESIS EVOLUTIONISTS

My narrative to this point has emphasized the alienation of embryologists from the MCTH and therefore from the Synthesis. This may have distorted the image of the early geneticists with respect to their views on development. From the modern standpoint of evo–devo, this is a tempting mistake: The absence of development from the Evolutionary Synthesis can be blamed on the absence of development from the MCTH, on which the Synthesis is based. The true story is more complex. There is no historical evidence to support the notion that the early MCTH advocates actually opposed the study of development. In fact, there is considerable evidence that they were very interested in it. Historian Robert Kohler claims that the Morgan group worked actively on developmental and evolutionary problems but failed to achieve publishable results. He blames the gulf between genetics and development on differences in experimental methods, and in the different experimental organisms of the two fields (Kohler 1993: 1061; Kohler 1994: 243). Many of the early geneticists, especially including T. H. Morgan, showed strong interest in development; they merely insisted that it be understood genetically. Morgan's early symbolism for genetic mutations has now been forgotten, but it reveals his concern for development. It was designed keep track of the entire number of genes that were known to developmentally influence a trait (Falk and Schwartz 1993). The notation proved too cumbersome and was abandoned. However, if it had been adopted, it would have been very difficult to point to one allele (p) and assert that it was "the cause" of an eye color when the genetic notation for eye color made mention of the other twenty-four loci known to affect it. (If only one of the twenty-five were the cause, why are all of the other noncauses even listed?) Even Hermann Muller, who was much more atomistic than Morgan regarding the relation between genes and characters (and possibly the author of the 1915 "genes cause characters" locution), had carefully thought out developmental–genetic ideas (Falk 1997).[51]

---

[51] In my early work on this topic I found it irresistible to conclude that the Morgan group was disinterested in embryological development. With patience and perseverance, Gar Allen and Raphael Falk have convinced me that I was wrong. The disinterest in development appeared not in the early geneticists, but within the Evolutionary Synthesis. The evolutionists' disinterest was *enabled* by the MCTH, but the founders of the MCTH did not share in it.

Even though Morgan and other early geneticists were themselves interested in the role of genes in ontogeny, they had made it *possible* to study heredity while ignoring development. When the Synthesis got underway, its practitioners did just that. Morgan himself was interested in development, but he could show no reason why evolutionists should be interested in it. During the nineteenth century it had been impossible to study heredity without studying development. Now it was possible, and the Synthesis architects did it. The most powerful spokesman of the Evolutionary Synthesis was Theodosius Dobzhansky – another student of Morgan. Transmission genetics was heredity, and it was at the absolute core of the new Synthesis. No reason could be seen to incorporate development in the Synthesis. The marginalization of development had been enabled by the MCTH. It was carried out within the Synthesis. Heredity must be included in an evolutionary theory, but why must development?

So even without active opposition to the inclusion of development within the early Evolutionary Synthesis, it was not included because it didn't have to be. In the following sections I discuss several examples of cooperation, or potential cooperation, among researchers that *might have* produced a developmental aspect to the Synthesis.

9.4.1 Sewall Wright

The most intriguing source of developmental interest is Sewall Wright. Wright was one of the inventors of mathematical population genetics (with R. A. Fisher and J. B. S. Haldane), and his collaboration with Dobzhansky during the 1930s and 1940s was a legendary strength of the growing Synthesis. Aside from population genetics itself, Wright's major influences on the Synthesis were his technique of representing variations in adaptation as a landscape, and his careful attention to the possibility of genetic drift in small populations. He was the only major figure within the Synthesis to study the physiological (and therefore the developmental) effects of genes.[52] His work was highly respected among the cytoplasmic geneticists of the 1940s and 1950s (and indeed was quoted in Ephrussi's aforementioned critique of Dobzhansky). Most geneticists of the Morgan group had claimed that any apparent influence of cytological factors on heredity could actually be accounted for by nuclear genes (Sapp 1987: 100). In contrast, Wright took the cytoplasmic

---

[52] He was hired by the University of Chicago in 1928 because of his interest in connecting genetics to development. Perhaps surprisingly, the hire was made by Frank Lillie (Provine 1986: 169). Not all embryological critics of genetics were so broadminded; Ross Harrison refused to hire a geneticist at Yale as late as the 1940s (Maienschein 1991b: 286).

effects seriously: "[T]here are cytoplasmic properties transmitted without apparent decay through many generations. ... Unless it is demonstrated that there is ultimate replacement by substances of nuclear origin, it is superfluous to trace them at all to nuclear genes" (Wright 1941: 501). The explanation of differentiation by means of persistent cytoplasmic changes (e.g., what were then called *plasmagenes*) is "the most probable view." But there is a catch. "The chief objection is that it ascribes enormous importance in cell lineages to a process which is only rarely responsible for differences between germ cells, at least within a species" (Wright 1941: 502). This remark implicitly acknowledges a Mendelian blind spot: If cytoplasmic characters differ only *between* species, the Mendelian could never know it! Nevertheless, Wright rejects the fundamental–superficial distinction on the basis of the discovery of early, mostly lethal mutations (Wright 1945: 301). So the paradox remains: Wright admits that cytoplasmic mechanisms are probably involved in differentiation, but genes are still regarded as the cause of differentiated characters on the grounds that they explain the differences between species members. If some solution to the problem of differentiation had been available to Wright in the 1940s, perhaps he would have found a way to incorporate it into the Synthesis. No such solution appeared. As far as I can judge, Wright's interest in development and the cytoplasm had no affect on his population genetics, or on the treatment of development within the Synthesis.

9.4.2   Oxford Morphology

A second potential source of structuralist–developmentalist influence on the Synthesis is the Oxford group of morphologists, E. S. Goodrich, Julian Huxley, and Gavin de Beer. Goodrich had been a student of E. Ray Lankester. He was an extremely prominent morphologist, and he articulated a possible compromise of Mendelism and Darwinism at the very early date of 1912, when the two theories seemed most opposed (Ruse 1996: 287). Huxley and de Beer were his students. Huxley was a remarkably broad theorist and an important participant in the Synthesis, even to the point of naming it in *Evolution: The Modern Synthesis* (Huxley 1942). Besides his evolutionary writings, Huxley was a student of allometry (correlations of growth patterns in different parts of the body) and coauthored an embryology text with de Beer. De Beer's little book of 1930, *Embryology and Evolution*, was revised and retitled *Embryos and Ancestors* (De Beer 1951) and is warmly regarded by recent evo–devo workers. Huxley certainly contributed to the Synthesis, but did he contribute embryology? Did Goodrich and de Beer contribute at all?

The two papers on morphology in Mayr and Provine's *The Evolutionary Synthesis* dismiss the importance of morphology to the Synthesis. Ghiselin believed that pre-Synthesis morphologists were too descriptive to contribute, and Coleman that they were too typological (Coleman 1980; Ghiselin 1980). A deeper insight comes from Churchill's paper in the same volume (Churchill 1980). He shows that Huxley did, but de Beer did not, fully understand the significance of the modern dichotomies: heredity versus development, germ versus soma, and genotype versus phenotype. Churchill exposes de Beer's misunderstanding of the Huxley–Goldschmidt concept of "rate genes." Huxley used the concept in an explanation of how the appearance of recapitulation could be caused by ordinary genetic factors, if these included genes that determine rates of physiological processes. De Beer used the same term, but in a subtly different manner. For de Beer, rate genes were not genes that specified various rates of developmental processes, but *genes that changed their rates!* Huxley thought in terms of gene frequencies in populations, and the fixed gene–trait correlations that count as causation under the MCTH. De Beer thought in terms of individual ontogenetic processes, and genes as epigenetic causal actors that can change their behavior. Huxley was a population thinker; de Beer was a developmental thinker (not to say a typological thinker). Churchill concludes his paper by reporting the remark of Ernst Mayr that "Huxley wasn't really an embryologist." Churchill had originally thought this comment unfair, but found it "curiously accurate" after recognizing the contrast with de Beer (Churchill 1980: 120).[53] "Not really an embryologist" meant "really a population thinker." Huxley was a population thinker who tried to introduce developmental concepts such as allometry and growth co-efficients into the Synthesis. The book that named the Synthesis states in its Preface that "a study of the effects of genes during development is as essential for an understanding of evolution as are the study of mutation and that of selection" (Huxley 1942: 8). Huxley made absolutely no headway convincing

---

[53] Churchill is among the clearest sighted of historians, but even he is under the influence of contemporary scientific paradigms. He reports a comment of de Beer's, typical of embryologists, that every body part comes *both* from heredity and acquisition (development). Churchill believes that this shows de Beer's confusion about genetics. "What about those genes, bound to the chromosomes in every nucleus, aren't they the result of 'inheritance alone'?" (Churchill 1980: 118). Churchill is correct that the strict genotype–phenotype distinction implies that genes are the product of "inheritance alone." However, de Beer (as an embryologist) may well not have accepted that view. Contrary to the common prejudice, Weismann did not: He regarded germ-line sequestration as an embryological phenomenon (Winther 2001). According to many modern thinkers, de Beer and Weismann were correct about this matter. Ontogenetic development, not inheritance alone, is causally responsible for the distribution of "genotokens" to the body's cells (van der Steen 1996; Buss 1987).

his Synthesis colleagues on this particular topic. I suggest that Goodrich and de Beer made no contribution at all.[54]

9.4.3 Waddington and Schmalhausen

If there was to be a developmental aspect of the Evolutionary Synthesis, it would surely have come from C. H. Waddington. Many of Waddington's concepts involving the role of development in the evolutionary process were anticipated or duplicated by I. I. Schmalhausen. However, Schmalhausen was isolated in the Soviet Union during the Lysenko years, and Waddington lived his life among Synthesis evolutionists. The theoretical differences between them are slight. Both discuss how natural selection can operate to change the ontogenetic processes by which adult phenotypes are created (consistent with the Causal Completeness Principle). Both emphasize the fact that a given genotype can produce different phenotypes in different environments. Until recently, Schmalhausen was mistakenly credited with inventing the concept of *norm of reaction*, which describes the environmental variation of phenotypes that can be produced by a given genotype.[55] Furthermore, both men proposed a mechanism by which selection on *the ontogenetic propensity to produce a trait* (i.e., on the norm of reaction) could result in a trait that was originally induced by environmental causes coming under internal genetic control. Schmalhausen called this "stabilizing selection"; Waddington called it "genetic assimilation."

Waddington was that rare embryologist who had been convinced of the truth of Mendelian genetics prior to his embryological education. He worked in Spemann's lab, and later in Morgan's lab in California. His first published book was on genetics, and it carried endorsements on its cover from Haldane, Huxley, and Muller (Waddington 1939). His second was a serious attempt to coordinate the work of Spemann and Morgan, perfectly titled *Organisers and Genes* (Waddington 1940). During the early 1930s, he had worked at Cambridge with the embryologists Joseph Needham and Jean Brachet, trying to identify the chemical nature of Spemann's organizer. It became increasingly clear that the chemical properties of an embryological inducer (e.g., the organizer) could not alone be credited with changes in the induced tissue, because a wide range of chemicals (including synthetic ones) could duplicate the effects of induction. The "competence" of the induced tissue

---

[54] This conclusion notwithstanding Waisbren (1988), who purports to show but does not show that the Oxford morphologists actually contributed something of morphological significance to the Synthesis (Love 2003).

[55] Sahotra Sarkar points out that the true source was Richard Wolterek in 1909 (Sarkar 1999).

was at least as importance as the chemical nature of the inducer. True to his genetics, Waddington attributed both the properties of the competent tissue and of the inducing substance to genes: "Since it is the genes which control the characters of the animal and its tissues, it must in general be the genes which determine the properties of the competence" (Waddington 1940: 54). Nevertheless, Waddington emphasized the ontogenetic processes that genes participate in, the "evocator-competence system." He continually advocated the need to include an understanding of ontogenetic processes within the Synthesis.

It is possible to find some influences of Waddington and Schmalhausen in mainstream Synthesis literature, but they are small (see Section 10.6 in the next chapter). Selection on details of ontogenetic processes is usually reinterpreted simply as background causes for increased adaptation. For example, developmental buffering will tend to increase adaptation in a partly stable environment (Mayr 1970: 108; Mayr's citations of Waddington will be discussed shortly). The norm of reaction was interpreted primarily in its adaptive meaning. In this way, environmentally caused variants could be presumed to be environmentally suitable and selection could be credited for this fact: "as far as possible, success in the past guarantees that each point on the norm of reaction is an adaptive reaction" (Wallace 1986: 160). This treatment removes two potential contributions of ontogeny to evolutionary change. One is that the norm of reaction would, under changed environmental circumstances, expose new phenotypic traits to selection, and these might (by genetic assimilation) become canalized in a species' phenotype. The other is that past genetic assimilation would result in the shielding from selective view of unexpressed genetic variation. This might later become unpredictably expressed and open to selection under changed environmental conditions. Wallace's adaptationist idealization of the norm of reaction as a wholly successful tuning of the genotype to every environmental contingency removes the need to carefully consider ontogeny as a causal factor in ongoing evolutionary changes.

Sahotra Sarkar shows that Dobzhansky even reinterpreted the norm of reaction to apply to populations rather than individuals. This completely removed its connection with ontogeny. "Dobzhansky, unlike Schmalhausen, and like a true geneticist from that period, generally ignored embryology" (Sarkar 1999: 246). The final irony came with the Synthesis reinterpretation of Schmalhausen's term for genetic assimilation, *stabilizing selection*. The meaning of this expression has completely changed within mainstream evolution discussions from Schmalhausen's intention. It now applies to selection for the mean in a population, as opposed to directional selection for extremes of a trait. This was not Schmalhausen's meaning. He had intended that *ontogenetic processes* were stabilized, and so buffered against either genetic or environmental perturbation. The Synthesized version of the expression removes all reference to ontogeny, and replaces it with a population–genetic definition of selection for the average phenotype. Semantic modifications such as these are *specific examples* of the black-boxing of embryology within the Synthesis. The black box is constructed out of population-level reinterpretations of concepts that were intended to refer to ontogenetic processes.

The work, especially of Waddington, deserves much more discussion than is possible here, but it seems clear that Waddington's hopes to significantly influence the Synthesis were not fulfilled. An indication of the suspicion with which he was held are Ernst Mayr's periodic suggestions that Waddington harbored Lamarckian leanings (Gilbert 1991: 205 n. 53). Further discussion of Waddington's interactions with the Synthesis will follow.

9.4.4   Richard Goldschmidt

Unlike most geneticists and embryologists, Goldschmidt considered genetics to be responsible for the explanation of ontogeny, and he tried to construct an evolutionary view that incorporated both. Knowing this alone, one might expect him to have been able to interest Synthesis authors in development. The effects were exactly the opposite. Goldschmidt had been among Germany's most prominent geneticists prior to his dismissal by the Nazis in 1935. He moved to the United States, and *The Material Basis of Evolution* was published in 1940 (Goldschmidt 1940). The book conflicted with the emerging evolutionary consensus in so many ways that it came to serve as a rallying point for the Synthesis architects. Goldschmidt denied gradualist views of speciation, the extrapolation of macroevolution from microevolution, and the particulate gene. The term *hopeful monster* was a mere throwaway expression for Goldschmidt, but it became the label by which neo-Darwinians stigmatized his theories. Michael Dietrich, in a fascinating study, has argued that Goldschmidt had a formative influence on the emerging Synthesis by providing a common foe to which the architects could react (Dietrich 1995). The actual episode is far too complex to discuss in the present context, but it almost seems possible that the Goldschmidt affair poisoned the waters of the Synthesis for the efforts of moderate developmentalists such as Waddington.

On the other hand, it would be naïve to assume that a coherent program could have been constructed in the 1950s that included both population genetics and developmental biology. (Or, I sometimes think, even today!) We have seen some factors that tend to devalue developmental contributions. One such factor is the strong resistance to cytoplasmic heredity. Another is the tendency to reinterpret developmental phenomena as populational in nature: Schmalhausen's stabilizing selection reinterpreted as acting merely on phenotypes rather than on ontogenies, Dobzhansky's populational redefinition of norm of reaction, and the claims by Dobzhansky and Mayr that Waddington's genetic assimilation experiments had only shown ordinary selection on threshold effects. As Dietrich reports, the Synthesis authors' decisions to target Goldschmidt for special criticism may have been a sociologically important event in unifying the movement. It is not clear how, or how much, Goldschmidt's incorporation of development into evolution was responsible for the fierceness of his rejection, but it surely played a role.

Nevertheless, there is little evidence of the *conscious* rejection of the relevance of development by Synthesis authors. Dobzhansky in 1951 had cited embryology alongside paleontology as one of those fields that had been proved *consistent* with neo-Darwinian population biology and therefore with the Extrapolation Principle. Attempts by structuralists to make development relevant to evolution were pretty regularly rebuffed, either gently (like the reinterpretations of Schmalhausen and Waddington) or harshly (like Goldschmidt).

The open antagonism between advocates of Synthesis evolutionary theory and structuralist evolutionists did not become public until two decades later. Between times, important methodological writings of around 1959 would intervene and set the stage. I discuss these in Chapter 10.

9.5 HISTORICAL REFLECTION: EXPLANATORY GOALS

At the end of Chapter 4 I suggested that the difference between Darwin and the morphologists was that they differed in their explanatory goals. The morphologists' goal was the explanation of form. Darwin's goal was the explanation of change. The morphological goal (form per se) required an understanding of ontogeny to support its phylogenetic conclusions. The Darwinian explanatory goal (changes, whether they be changes in form or in any other trait) did not require an understanding of ontogeny. This chapter's sketch of the contrast between Evolutionary Synthesis theorists and their embryological critics shows this same contrast.

9.5.1 *Form-Theoretic Evolutionary Theory*

Understanding form in an evolutionary sense requires understanding the ontogenetic processes by which form is produced, and the ways in which those ontogenetic processes can be changed during evolution. This gives a deep understanding of the production and change of form, and the relationships among organisms that are due to shared developmental processes. (Lillie: a system that cannot explain ontogeny a fortiori cannot explain phylogeny.) Shared aspects of developmental processes give rise to homologies. These aspects are genuine, causally active universals in the current world, and they help us understand the unity of life.

9.5.2 Change-Theoretic Evolutionary Theory

Understanding evolutionary change requires (and *only* requires) understanding the processes by which ancestral populations of interbreeding organisms give rise to descendant populations that have different heritable characteristics. According to the Evolutionary Synthesis, these processes involve mutation, selection, migration, and drift. The evolving traits may be traits for which the ontogeny is understood, like morphological characters, but they may equally well be traits of unknown ontogeny, like behavior or instinct. Ontological origins are irrelevant to heredity, and so the ontogeny of a character is irrelevant to the evolutionary explanation of its change. Any structuralist perception of universals of ontogeny is a conceptual confusion that distracts from the recognition of the populational nature of the process. Homologies, for example, are mere by-products of history, not indicators of ongoing unities.

10

The Synthesis Matures

The 1950s were a period of consolidation for the Evolutionary Synthesis. Or a period of increasing self-awareness as a movement...or a period during which those who wished to present evolutionary biology a cohesive body of scientific knowledge were more successful in doing so. A conference held in Princeton in 1947 had resulted in an unusual amount of agreement on theoretical issues, and this encouraged the feeling of unity (Davis 1949). However, a note of discord appeared in an Oxford Symposium on evolution in 1951. Waddington gave a paper entitled "Epigenetics and Evolution." He said that the achievements of "mathematical theorists on the one hand and experimental naturalists on the other" had been so striking that evolutionary scientists might have been seen to have "reached their goal with some degree of finality" (Waddington 1953: 186). Not so, said Waddington. The achievements of mathematical and experimental geneticists were less impressive to him. Embryologists, including Goldschmidt, Schmalhausen, and Dalcq, had continued to raise questions. The process of ontogenetic development had been neglected, and without it the Synthesis must remain incomplete. Waddington called attention to the dichotomy of genotype and phenotype, and he claimed that it ignored the *epigenotype*, Waddington's term for the processes of ontogeny. He explained why the epigenotype must be included in any evolutionary understanding:

> Changes in genotypes only have ostensible effects in evolution if they bring with them alterations in the epigenetic processes by which phenotypes come into being; and the kinds of change possible in the adult of any animal are limited to the possible alterations in the epigenetic system by which it is produced. (Waddington 1953: 190)

This public announcement of the Causal Completeness Principle may have been the first to take place at an openly Synthesis-oriented session. Waddington was clearly a forerunner of evo–devo, but it would be overly simplistic to track the Synthesis-versus-development conflicts to Waddington 1953. The reason is this: Certain theoretical, historical, and philosophical formulations of Synthesis biology were crucial to the exclusion of development in the 1970s and later. Those formulations were not available in the early 1950s, because they were devised by Ernst Mayr at the end of that decade. In fact, Waddington's protest was almost completely ignored by mainstream evolutionists. It received only one response, and that had to wait until 1959 (Provine 1980: 402). Protests like Waddington's would reappear in the late 1970s and 1980s. The attacks would be more vigorous, and so would the responses. By that time, the conceptual repertoire of the Synthesis had been expanded in ways that made the rejection of development much simpler. This expansion was largely the result of Ernst Mayr's "flurry of articles" on methodological topics published around the time of the Darwin centennial celebration in 1959.[56] As we will see in Chapter 11, by 1980 the Synthesis proponents had clear methodological grounds on which to reject the relevance of development.

Anyone familiar with evolutionary biology knows that Ernst Mayr is a unique figure. His organizational skills and energy were important to the early formation of the Society for the Study of Evolution in 1946; his science has been vitally important for the Synthesis; his scientific popularization and his outreach from evolutionary to other areas of science are important; and (unlike many other broadly active scientists) he has been extremely influential in the modern subdisciplines both of philosophy of biology and history of biology (Grene and Ruse 1994). However, one aspect of the formation of the Evolutionary Synthesis has not yet been discussed. It is the conflict, much older than the Synthesis, between the experimentalist and naturalist traditions in science. As the Synthesis was consolidating (or whatever it was doing) in the 1950s, traditional naturalistic biology (*organismic biology*, as it came to be called) was under challenge from the growing power of experimentalism and molecularization. This was taking place on the direct practical grounds of funding for research and academic positions (Beatty 1994: 351). Those of us who like to think of the Evolutionary Synthesis as the major biological achievement of the twentieth century should be aware that the rest of the

---

[56] Reference to the "flurry of articles" is from Chung 2003; it seems appropriate. Crucial philosophical and historical concepts came thick and fast. They shaped evolutionary thought up to the present day.

world considers it to be the discovery of the molecular structure of DNA by Crick and Watson in 1953.

Mayr came to recognize the molecularization of biology as a serious challenge to the naturalist tradition. The opportunity arose to present the Evolutionary Synthesis itself an achievement of naturalistic science, but to do so meant to assert the importance of field naturalists rather than mathematical geneticists in the formation of the Synthesis. This was a factor behind Mayr's interest in historical and philosophical concepts, and his activities with the upcoming centennial of the publication of Darwin's *Origin of Species*. The celebration would be an opportunity to align the Synthesis with Darwin, to educate the public about modern evolutionary theory, and thereby to enhance recognition and public status especially of the naturalistic (as opposed to molecular) aspects of evolution. Mayr wanted to challenge the importance of the mathematical genetic basis of the Synthesis so that the Synthesis itself could be a bulwark of naturalistic science against the rising tide of molecularization.

The American celebration of the 1959 Darwin centennial took place at the University of Chicago. Its timing, twelve years after the Princeton conference, was ideal for a celebration of the unification of evolution theory. The occasion was used to "reinvent [Darwin] as the 'founding father' of their discipline" (Smocovitis 1999: 279). The organizers included Sol Tax, who wanted to bring an evolutionary approach to his field of anthropology, and the results were published in a series of three volumes (including Tax 1960). The organizers intended to give a broad perspective on evolution, and they proposed the inclusion of papers on social evolution and possibly even extraterrestrial evolution. The centennial plans were coordinated with the major figures of evolutionary biology, many of whom were conveniently gathered at Cold Spring Harbor Laboratories for the weeklong symposium that began on June third. Ernst Mayr's "Where Are We?," one of the papers of his 1959 flurry, was the first paper read at that conference (Mayr 1959c). Smocovitis has located a letter written from Cold Spring Harbor to the centennial organizers regarding the gathered evolutionists' opinions about the proposed centennial program. It was written by Alfred E. Emerson, a prominent entomologist and speciation theorist, and dated June 7, 1959. Emerson reported that he had conferred informally with Sewall Wright, Dobzhansky, Rensch, Mayr, and Stebbins about the proposed program, and reported their opinions. According to Emerson, the evolutionists had disapproved of speculations about life on other planets, and looked askance at certain other topics. "Phases of biology, no matter how important, not associated with modern evolutionary thought (*i.e., development of the living organism* or the contemporary society) should not be emphasized in a Darwin centennial" (Smocovitis 1999: 296; emphasis added). So much for Waddington.

The confluence of the Darwin centennial, the pressing need to defend naturalistic biological sciences, and his increasingly prominent role as spokesperson for evolutionary biology seem to have inspired Ernst Mayr to explore the historical and philosophical roots of evolutionary biology. Mayr articulated two important dichotomies around 1959. Although each of them has a history in Mayr's own intellectual development, the two dichotomies only took on a truly universal form around the time of the centennial. They were remarkably successful in legitimating naturalistic studies against the incursion from molecularization, and in reinforcing the conceptual framework of the Synthesis. Each doctrine became embedded in the humanistic fields of history and philosophy of biology, and each of these fields in turn began to flourish in part because of Mayr's interest in them. Each doctrine would later play an important role in the structuralist debates against the Synthesis.

10.2   USES OF DICHOTOMIES

Mayr's two new dichotomies were *proximate versus ultimate causation*, and *population thinking versus typological thinking*. Before we consider the context of origin and the consequences of these dichotomies, let us review two older examples. After the split between heredity and development, advocates of the importance of development to evolution had been on the defensive. The structuralists' Causal Completeness Principle had been a nearly universal assumption during the nineteenth century. Heredity was an aspect of development, heredity was involved in evolution, and therefore evolution was to be understood as changes in developmental processes. The Mendelian–chromosomal theory drove a wedge between heredity and development. Genes were said to be the *causes* of organic (even adult) traits. This new sense of cause allowed heredity to explain traits at a distance, so to speak – without tracing a continuous causal path through ontogeny. The pinkness of the fly's eye was *embryologically caused* by chemical and cellular actions during its development. No one denied this. However, it was *hereditarily caused* merely by the fly's possession of a particular gene. Therefore, the dichotomy between heredity and development took embryology out of the game of hereditary causation.

Back when heredity was an aspect of development, evolution was a change in the developmental process by which bodies are built. With the new, narrow concept of heredity, body traits were caused directly by genes. So evolution became *changes in genes*, rather than changes in developmental processes. Development was left behind, or rather it was black boxed, set aside, and ignored. If development really is relevant to evolution, some argument must be constructed to prove its relevance.

With the acceptance of the MCTH and eventually the Synthesis, the vocabulary of biology began to be standardized in ways that fit the modern scientific ontology. We have already discussed two historical dichotomies that were reinterpreted for modern use. One was Weismann's germ–soma distinction, and the other was Johannsen's genotype–phenotype distinction. We discussed in Chapter 7, Section 7.5 how these distinctions have taken on a meaning quite distinct from that intended by their inventors. The new meaning is aligned with a doctrine that is implicit in the MCTH and especially the Synthesis: Organisms have exactly two scientifically important aspects, and one is the cause of the other. One aspect is the observable body, and the other is the hereditary cause of the observable body. The observable body is the phenotype; the hereditary cause is the genotype. The embryological causes that intervene between genotype and phenotype are irrelevant to the study of either heredity or evolution. Embryological phenomena do not appear as an element in binaries like genotype–phenotype. Alternative theories are put at a disadvantage by these dichotomies. Sometimes the advocates of alternative theories try to expand the dichotomies. Waddington's 1953 paper renewed his arguments from as early as 1939 that attention must be paid to "epigenetics" and the "epigenotype," the kind of causation that intervened between genome and phenome. He had little success.

Neither the proximate–ultimate nor the population–typology dichotomy was invented in order to deal with challenges from developmental evolutionists. The proximate–ultimate distinction has a respectable history in jurisprudence, and Mayr had been familiar with its biological use by naturalists such as David Lack (Beatty 1994). The population–typology dichotomy grew out of Mayr's ongoing attempts, since at least 1942, to develop a theoretically adequate concept of species and to account for the historical differences among older species concepts (Chung 2003). Both distinctions came to fruition around the time of the centennial. Both served the purpose of explaining the unique importance of naturalistic (as opposed to molecular and experimental) scientific studies. Molecular studies of biology, such as those of Crick and Watson, deal only with proximate causation. To understand ultimate (evolutionary) causes, one must study evolution, and study it naturalistically. The relevance of the population–typology distinction to the naturalist-versus-experimentalist conflict is subtler. Mayr needed to resist the impression that mathematical geneticists and experimentalists were responsible for the Evolutionary Synthesis. He wanted to show that real naturalists (not abstract mathematical theorists) had discovered the populational nature of species from naturalistic observations in the field. He did this by discovering population thinking among earlier naturalists, who predated the dawn of genetics: "The claim has been made by some population geneticists that population thinking and its application to evolutionary theory is a contribution of genetics. This overlooks that population thinking was already strongly apparent in Darwin's own work" (Mayr 1959c: 3).

Smocovitis's assertion that Darwin was "reinvented as the founding father" of the Synthesis at the centennial may sound extravagant, but I think it is quite accurate. As Gould has pointed out, the Synthesis "hardened" and became more adaptationist between its inception and 1959 (Gould 1980, 1983). Even though Gould disapproves of the hardening, it was done in response to the available empirical evidence that drift was a less important parameter than selection. Thus, the Synthesis really *was* more Darwinian in 1959 than 1945, at least in the sense that selection was a more important aspect of the theory. In addition, many people today think of Darwin as the originator of population thinking, and population thinking as the core of the Synthesis. Population thinking itself was around long before Mayr named it and contrasted it with typological thinking (in Mayr 1959b). But when was Darwin recognized as its inventor? Mayr's same 1959 publication, the one that named the contrast, was the first to finger Darwin as the founder of population thinking. Mayr had discussed the history of populational concepts in earlier writings. In 1953 he said that the population concept of species had gradually been replacing the type concept since 1878 (with no reason given for the date). In 1955 he identified a 1905 paper by Karl Jordan as an important influence on population thinking. Not until 1959 did Mayr's publish his vision of the history of population thinking stretching all the way back to Darwin (Chung 2003). Darwin really was reinvented as a population thinker, and for that reason he was named the forefather of the Synthesis. His parentage was established on the centennial of his major publication.

10.3 PROXIMATE VERSUS ULTIMATE: CONTEXT

Mayr introduced the proximate–ultimate distinction in "Cause and Effect in Biology" (Mayr 1961). Although it deals with the concept of causation, it can be equally well understood in terms of kinds of explanations, which I prefer. It illustrates (but does not name) the principle of explanatory relativity. When presuppositions differ, it is possible that (what sounds like) the *very same fact* is explained by apparently inconsistent explanations (see Chapter1, Section1.6). The contrast between proximate and ultimate causation is a careful and specific recognition of the relativity of explanation. What looks like the same fact can be given either a proximate or an ultimate explanation. Recognizing the difference between these two kinds of explanation can help us to avoid needless controversy.

That's the good side of the proximate–ultimate distinction. The bad side is that the application of the distinction can be just as biased as any other bit of scientific rhetoric. For example, why is the distinction binary? Couldn't the contrast be a matter of degree, with ontogeny conceived as an intermediary stage between ancestral selection and an adult trait? Mayr later discussed the difference between closed and open "behavioral programs," and he even discussed the ontogeny of behavior in open programs. Open programs are those in which a significant amount of learning occurs (Mayr 1974). An adult's response to a stimulus is clearly a matter of proximate causation. However, the adult's behavior is only partly determined by the ultimate evolutionary origin of its genotype. It is also determined in part by the environmental influences on the adult during its early development. The influences from early ontogeny are clearly "more ultimate" than its present stimulus, but not so ultimate as the evolutionary selection for its genotype. Why not conceive of ontogeny as an in-between point in the proximate–ultimate scale?

This would seem to be consistent both with Mayr's original definition of the contrast and with his discussion of behavior. It would then enable other kinds of ontogeny (e.g., Waddington's epigenotype) to be conceived as partly ultimate, and therefore relevant to the understanding of evolution. In fact, though, this is not the direction Mayr took. The proximate–ultimate distinction remained binary. Eventually the binary nature of the distinction would be used against structuralist biology. Details will follow.

In 1950, a Synthesis evolutionist at a Cold Spring Harbor Symposium presented a paper with a section entitled "Race as a Type and as a Population." It contrasted the modern populational concept of species with the view that species were types, and it associated the type-view with pre-Darwinian thinking and "Platonic immutable ideas that are only imperfectly manifested in the world." It went on to claim that racism is based on type thinking, and that population thinking is the cure. The author was not Ernst Mayr, but Theodosius Dobzhansky (Dobzhansky 1950).

Ernst Mayr is nevertheless indelibly associated with the contrast between population thinking and typological thinking. This is quite appropriate. The immense effect this distinction has had on science, history, philosophy, and even popular culture stemmed not from Dobzhansky but from Mayr. Dobzhansky made a vague suggestion that Platonic typology might have been attractive in pre-Darwinian times, but he made no historical assertion to that effect. The individuals he labeled as type-thinkers were merely some old-fashioned taxonomists, and (of course) modern racists. Dobzhansky had labeled both pre-Darwin species fixists and modern racists as flawed thinkers on grounds of their typology. Mayr was to continue this dual use of typology, as explanation both of past errors and modern confusions.

Carl Chung's study of the development of the population–typology distinction in Mayr's thought shows that it began with concerns about the proper definition of the species category, and how best to understand the species concepts of past eras (Chung 2003). Precursors of the population and typology concepts were the species concepts associated with "new systematics" and "old systematics" as discussed in Chapter 8, Section 8.5. Chung shows that, between 1942 and 1953, Mayr treated the topic as a restricted taxonomic issue, not something of general biological importance. However, in the 1955 paper that identified Karl Jordan as a population thinker, the population–typology distinction is elevated to a matter of concern to all of biology. In 1959 it was elevated again, to become a major theme in the entire history of Western thought:

> Most of the great philosophers of the 17th, 18th, and 19th centuries were influenced by the philosophy of Plato, and the thinking of this school dominated the period. Since there is no gradation between types, gradual evolution is basically a logical impossibility for the typologist. Evolution, if it occurs at all, has to proceed by jumps or steps. (Mayr 1959b: 2)

Mayr immediately recognized the forcefulness of this contrast, and it became a central theme of his writing. He quoted his own 1959 passage in his preface to the 1964 Harvard University Press facsimile reproduction of Darwin's *Origin*, in his 1966 *Animal Species and Evolution*, and elsewhere.

Ancient Greece was in the air, and two other authors stated similar dichotomies at about the same time. One was A.J. Cain, who explained Linnaeus's species fixism as following from ancient Greek philosophy (Cain 1958). The other was philosophy graduate student David Hull, in a paper published as "The Effect of Essentialism on Taxonomy: 2000 Years of Stasis"

(Hull 1965). Some tensions existed in the fact that Cain and Hull blamed Aristotelian essentialism for the evils, whereas Mayr had blamed Platonic idealist Types. In 1968 Mayr made the historical accommodation:

> Essentialism considers it the task of pure knowledge to discover the hidden nature (or form, or essence) of things. When applied to organic diversity, it believes that all members of a taxon share the same essential nature; they conform to the same type. This is why essentialist ideology is also referred to as typology. Classification of organic diversity of the essentialists consists in assigning the variability of nature to a fixed number of basic types at various levels. Variation is considered a trivial and irrelevant phenomenon. (Mayr 1976: 248)

Mayr and Cain had originally been very expansive in their attributions of essentialist–typological influence to historical periods. Mayr had taken it through the nineteenth century, and Cain claimed that the abandonment of a priori classifications in favor of empirical ones "was not complete when the theory of evolution arrived" (Cain 1958: 147). In 1968, Mayr backed off a bit and recognized that the years between Linnaeus and Darwin were "a period of transition" to empirical methods, although typological influences still existed. Hull's title sounds even more expansive: 2000 years of stasis in systematics! The title matches a frequent slogan of nominalists, one that Hull quotes from Karl Popper himself:

> [E]very discipline as long as it used the Aristotelian method of definition has remained arrested in a state of empty verbiage and barren scholasticism, and that the degree to which various sciences have been able to make any progress depended on the degree to which they have been able to get rid of this essentialist method. (Hull 1965: 314; also Popper 1950: 206).

Hull accepted Cain's claim that Aristotle had bewitched pre-Darwinian taxonomists. Contrary to his title, though, he did not address essentialism regarding the definition of species taxa or of higher taxa. Instead he discussed the contemporary debates among "new" systematists about the definition of the species category (whether it should be defined in terms of morphology, sterility, etc.). Hull argued that just as individual taxa cannot receive Aristotelian necessary-and-sufficient definitions, scientific concepts such as *species* should not be expected to either.[57] Nevertheless, his articulation of

---

[57] The peculiar mismatch between title and content occurred because Popper had submitted the paper for publication without informing Hull. Hull did not fully endorse the original paper's content, which had been designed to please Popper, his instructor. He revised the content to its nonnominalist form but left the nominalist title (Hull, personal communication,

the doctrine of essentialism smoothed the unification of typology and essentialism in Mayr's mind.

Cain, Mayr, and Hull were soon joined by Michael Ghiselin, who in 1969 again cited essentialism as Darwin's primary foe (Ghiselin 1969: 50 ff.). This convergence marked the origin of the Essentialism Story, discussed earlier in this book in its historical application. In a sort of time-travel paradox, we have now seen the origins of the doctrine that we have struggled with throughout the present book. The quick acceptance of the doctrine was perhaps due to the fact that there were few professional historians of biology around in 1960, and Mayr knew it. When Gertrude Himmelfarb's biography of Darwin came out in 1959, Mayr reviewed it immediately and sternly. "To me it seems a major gap in Dr. Himmelfarb's presentation that she nowhere discusses the enormous impetus Darwin gave to 'population thinking' and the mortal wound he inflicted upon 'typological thinking'" (Mayr 1959a: 215). The very first description of Darwin as a population thinker was published that same year by Mayr himself, so it is not surprising that Himmelfarb was unaware of it.

Mayr later became extremely well read in primary sources in the history of biology, but no citations and no quotations from Darwin support the historical claims that were made in introducing the typology–population distinction. The historical research supporting his 1959 study of Agassiz (Mayr 1976) is acknowledged to be indebted primarily to one book, Arthur Lovejoy's *The Great Chain of Being* (1936). Lovejoy does express the notion that philosophical ideas can have effects that last unnoticed for millennia, but the philosophical concept Lovejoy stresses is not the one Mayr stresses. The title of Lovejoy's book refers not to the ancient Greek doctrine of the distinctness of kinds (typology–essentialism), but to its opposite. The Great Chain of Being was the scala natura, the linear arrangement of all kinds in a scale from lowest to highest. An essential feature of the scala was the principle of plenitude: Nature had no gaps. Lovejoy did acknowledge the contrary doctrine of distinct kinds, but he considered it less important. Mayr decided to emphasize the distinctness doctrine and almost ignored the doctrine of continuity. Historian Mary Winsor has reported on an unpublished letter in which Mayr explains his dislike for Plato as stemming from his early education in Germany. In a 1956 letter to Carl Epling, Mayr reports that "Plato, under whose influence I have suffered throughout my high school and college career" was one of

August 7, 2002). The title echoes an old nominalist mantra. Jeremy Bentham in 1817 complained that "for little less than two thousand years, the followers of Aristotle kept *art* and *science* nearly at a stand" by "fancying that everything could be done, by putting together a parcel of phrases, expressive of the respective imports of certain *words*" (Bentham 1969: 280).

his "favorite scapegoats... It is good for my liver if I am permitted to knock [him] down from time to time." He insists to Epling that the theories of Goldschmidt, Schindewolf, and other German morphologists were clearly influenced by Platonic typology.[58]

Mayr's study of Agassiz confirmed his notion that pre-Darwinians were typologists and metaphysical idealists, and that they considered taxa to be ideas in God's mind. However, we have seen that Agassiz was an extremely unusual figure in this regard. The Essentialism Story fits no other pre-Darwinian author like it fits Agassiz. Even so, the early typological version of the Essentialism Story was conceived in the context of Goldschmidt and Schindewolf, not Agassiz. So when Mayr reported that Darwin had refuted typology in favor of population thinking, two messages were being sent. One was a historical claim about Darwin and population thinking, and the other was a contemporary claim about the *reasons for superiority* of Synthesis biology over its competitors. Philosophers and historians universally took the historical claim as accurate and authoritative. The accuracy was first historically challenged thirty years later, and then by an anthropologist (Atran 1990)!

In hindsight, it is embarrassing how very slim the historical evidence was for the Essentialism Story (Winsor 2003; also see Winsor forthcoming and Amundson forthcoming). Quotations that merely demonstrated an author's commitment to species fixism were taken as proof of the essentialist underpinnings of this belief. An example is Mayr's commentary on a quotation from Charles Lyell. Here is Lyell:

> It is idle... to dispute about the abstract possibility of the conversion of one species into another, when there are known causes, so much more active in their nature, which must always intervene and prevent the actual accomplishment of such conversions. (Lyell 1835 v. 3: 162)

Here is Mayr's commentary: "For an essentialist there can be no evolution, there can only be a sudden origin of a new essence by a major mutation or saltation" (Mayr 1988: 172). However, even in the quoted passage, Lyell had spoken of "known causes... active in their nature," not of metaphysical essences. Lyell dedicated the first four chapters of Volume 2 of the first edition of this work to evidence against Lamarckian transmutation (Lyell 1832). Many of these arguments were quite cogent, and none referred to essences. The universal acceptance of the Essentialism Story did not derive from the evidence in its favor.

---

[58] Quotations from the 1956 Mayr letter are from a paper in progress by Mary P. Winsor (Winsor forthcoming). Other interpretations in this section are also influenced by that paper.

An additional shortcoming of the Essentialism Story should have been noticed by philosophers, but apparently was not. Essentialism is a doctrine about natural kinds, not about the causal relations between these kinds. Its paradigmatic application is to items like geometric figures: A triangle cannot change into a square because their essences are distinct. In contrast, species fixism is a doctrine about causal relations – the causal relation of *generation* between parents and offspring. Essentialism may entail that a dog cannot transform into a cat, but it cannot (by itself) entail that a dog cannot *give birth to* a cat. Generation was a scientific problem of great significance in the eighteenth century, and it was important even to Darwin (Hodge 1985). Given the central importance of generation to the question of species fixism, essentialism was far too simple a doctrine to have played the role that it is said to have played (Amundson forthcoming).

10.5   ERNST MAYR AS A STRUCTURALIST?

The simplest stories are the best. In Chapter 11 I try to show how Mayr's two 1959 dichotomies came to be used alongside the older dichotomies of genotype–phenotype and germ line–soma as a barrier against the relevance of development for evolution. As we have seen in this chapter, Mayr was aware as early as 1956 that typology could be used to condemn not only pre-Darwinians but also the perceived opponents of the Evolutionary Synthesis, of whom Goldschmidt was the archetype. Nevertheless, during this period, Mayr did not place all advocates of the importance of development in the category of "opponents of the Evolutionary Synthesis." In fact, in a certain specific context, Mayr himself argued for the importance of development.

I have already reported that only one Synthesis figure responded to Waddington's 1953 protest. The single commentator was Ernst Mayr himself, and the response occurred in his important introductory address called "Where Are We Now?" given to the Cold Spring Harbor symposium on June 3, 1959 (Mayr 1959c). The surprise may be that Mayr supported Waddington's complaint! "With geneticists all around, Mayr explicitly challenged the view that geneticists, especially Fisher, Haldane, and Wright, were entirely responsible for the evolutionary synthesis" (Provine 1980: 402). This was in line with Mayr's strategy of identifying the Synthesis with naturalistic systematic studies instead of mathematical and experimental science. Mayr, however, went farther than that. He argued for a change in the target of selection, away from genes and gene frequencies and toward the phenotypes of individual organisms. Naturalists (he said) could see that entire phenotypes were the units on which selection operated. The early geneticists' technique of visualizing genes as representations of characters was outmoded. Mayr's first published attribution of typological thinking to modern scientists was made in that paper. It was applied to the saltationist Mendelians Hugo de Vries and William Bateson, and later in the paper to geneticists who had been overly influenced by the concept of the unit character. Mayr went on to cite positively Waddington's work on genetic assimilation.

Soon thereafter, in his monumental 800-page *Animal Species and Evolution*, Mayr began to credit Waddington even more directly and to use his technical terminology of *epigenetic* and *epigenotype*:

> Our ideas on the relation between gene and character have been thoroughly revised and the phenotype is more and more considered not as a mosaic of individual gene-controlled characters but as the joint product of a complex interacting system, the total epigenotype [he cites Waddington on the epigenotype]. (Mayr 1966: 6; see also 148, 185, etc.)

Waddington is a major player in this book. In terms of numbers of index references, only Mayr's co-architects (Simpson, Dobzhansky, Stebbins, and Haldane) and the naturalists Lack and White receive more references. By this measure Waddington is more important to Mayr than Sewall Wright, Rensch, Muller, and Fisher. Mayr later reported that naturalists like him had only "adopted temporarily the absurd reductionist definition of evolution as 'a change of gene frequencies in populations'" (Mayr 1984: 1261). Mayr's transition from gene-selectionist to individual selectionist was greatly assisted by the phenotypic integration he took to be implied by Waddington's developmental approach.

Mayr's attraction to Waddington's developmental explanations was somewhat superficial. As Scott Gilbert pointed out, he did not accept Waddington's account of genetic assimilation, and he attributed it to ordinary selection on simple threshold effects instead of complex developmental processes. However, during the time Mayr was arguing against the reductionism of the gene-selectionists, Waddington was treated as an important ally. One wonders about the letter of June 7, 1959, sent by Emerson from Cold Spring Harbor Laboratories. Had Mayr, four days after the "Where Are We Now?" talk in which he sided with Waddington, actually concurred with his colleagues that developmental biologists were "not associated with modern evolutionary thought"? Mayr was apparently willing to use Waddington's arguments as grounds to favor individual selection over genic selection, but *not* as grounds to consider developmental thought relevant to evolutionary theory.[59]

During the 1970s a movement sprang up that criticized the Evolutionary Synthesis, partly on grounds of its failure to include development. This time Mayr stood with the Synthesis against the critics. Waddington's approach was no longer useful, and Mayr began to treat him more skeptically. In addition, he and others began to use the two new dichotomies to defend the Synthesis against its critics. The details will be treated in Chapter 11.

Before we leave the period of the centennial, one more of Mayr's innovations should be acknowledged. Evelyn Fox Keller has studied the history of the various metaphors that have been used to describe the relation between gene and organism. In a paper on the origin of the *genetic program* concept, she identifies two papers as independent origins of the term *program* in reference to genetics. Both papers were published in 1961 (Keller 2000b: 176 n. 7). Jacques Monod and François Jacob presented one at Cold Spring Harbor, the same year as their groundbreaking paper on the operon concept in bacterial adaptation, a paper widely reported to have resolved the Developmental Paradox. The other was Ernst Mayr's "Cause and Effect in Biology," the modern source of the proximate–ultimate distinction (Mayr 1961). The genetic program concept was not entirely fruitful for an epigenetic developmental perspective, of course. It continued to attribute causal responsibility to the genome rather than to epigenesis. However, it again shows Mayr's contribution to the cutting edge of conceptual activity of this period.

10.6   THE ENLARGED QUIVER OF DICHOTOMIES

The two new dichotomies produced by Ernst Mayr around 1959 were not designed to refute any claims for the relevance of development for evolutionary understanding. After the protests of the 1970s, however, they were integrated into the Synthesis defenses against structuralist critics. The fact that the dichotomies had not been invented for this purpose may simply indicate that Synthesis authors had not planned from the start to keep developmental thinkers out of their new theory. This suggests that there really is something

---

[59] A second bit of evidence that Mayr only taking Waddington partly seriously is his statement that condemns "Any author who uses findings from the ontogeny of an individual to prove one or another evolutionary theory" (Mayr 1959b: 8). I do not feel that this quotation is best understood as a direct rejection of developmentalist thought, for reasons discussed in Chapter 11, note 63.

in the structure of the opposing theories that is at odds – the problem does not arise from personal motivations to undercut opponents.

To review, the four dichotomies are these:

- Genotype versus phenotype: This is the basic ontology of the Mendelian–chromosomal theory and the Synthesis. This dichotomy is treated as exhaustive for the purposes of mid-century genetics and evolution theory. The embryological processes that form the bridge between the zygote and adult are black boxed and ignored.
- Germ line versus soma: This was conceived as a developmental and embryological distinction by Weismann. It was reconceived as an ontological distinction that guarantees the separation of heredity from development (examples to follow).
- Proximate versus ultimate: This delineates two separate kinds of biological questions. It was used after 1970 to categorize developmental processes as proximate, and so to label them as logically irrelevant to ultimate evolutionary explanations.
- Typological thinking versus population thinking: Population thinking is a central aspect of Synthesis biology, and typology was conceived as its opposite. Typology (or essentialism) was never defined clearly enough to operationally identify its advocates except trivially, by their apparent opposition to population thinking.

We will now observe the effects of these dichotomies, and other methodological differences, in the debates about the Synthesis from around 1970 to the present.

11

Recent Debates and the Continuing Tension

11.1 DIVERSITY VERSUS COMMONALITY: STARTING WITH GENES

The adaptation-versus-constraint debates that began in the late 1970s were a part of a larger and more diffuse critique of the Evolutionary Synthesis. Mainstream neo-Darwinism was subject to a very wide range of methodological criticisms in those days, including the alleged unfalsifiability of adaptationism, the failure to consider nonselective evolutionary phenomena such as drift, and the inability to explain punctuation in the fossil record. The debate was also influenced by political factors, such as the association of adaptationist theorizing with status quo conservative politics (an association at least as old as the British natural theologians; see Desmond 1989). The arguments discussed in the following paragraphs are products of these debates. My own sympathy with the constraint side of these debates is already obvious to the reader. However, it would certainly be a mistake to accept at face value these critics' accounts of the shortcomings of Synthesis theory. We are searching for the genuine theoretical grounds for conflict between structuralist and adaptationist theories, and those grounds were often misstated in those debates. For that reason, this section examines a topic that preceded the constraints debates and that does not directly involve development at all. It exposes the incorrectness of at least one of the early criticisms of the Synthesis, but it also reveals a methodological tendency in Synthesis thought. The tendency was innocuous at first, but it played a contentious role in the later Synthesis stance against the importance of development. The topic is the notion of homologous genes and its treatment by Synthesis architects Dobzhansky and Mayr.

Natural selection reduces genetic diversity within a population. However, when coupled with speciation, it produces genetic and phenotypic diversity between populations. *Adaptive radiation* is the primary phyletic effect of evolution by natural selection, but diversity doesn't always increase. Sometimes

213

new similarities arise – homoplastic traits. These can be explained by natural selection also. Selectionist explanations of homoplasies always treat them as a secondary, coincidental effect that followed an earlier history of adaptive radiation. The selectionist explanation of convergence contrasts with possible developmental explanations, as we will see, but it also contrasts with a phenomenon that need not refer to development at all – the sharing of homologous genes among remotely related species. Similarities between remotely related species might be the result of the sharing of homologous genes even if the genes are understood in transmission genetic terms – the "gene for" the trait in each species is a descendant of the same gene in a common ancestor. Dobzhansky and Mayr were both skeptical about explanations in terms of homologous genes, and their reasons are related to the reasons for opposing developmental theories of evolution. However, because the concept of homologous genes carries no taint of idealism or typology, the grounds for resisting it can be recognized as distinct from the opposition to developmental theories in general. The reasoning behind this tendency gives an insight into the later development of specifically antidevelopmental arguments during the 1980s.

We have seen that neo-Darwinians had a faith in the sufficiency of transmission genetics that was not shared by embryologists of the period. Synthesis evolutionists were worried about neither the Mendelian blind spots nor the Developmental Paradox (Chapter 9, Sections 9.3.2 and 9.3.4). From a structuralist point of view, these barriers must be removed before any evolutionary theory could be presumed complete. Neo-Darwinians had no such worry. One possible interpretation of their complacency is that it was caused by an atomistic understanding of the gene-trait relationship. If adaptationists believed that bodily traits were independent of each other, controlled by distinct genes, adaptive convergence would be easier to achieve than if traits were developmentally and genetically linked to other traits. The accusation of atomism about traits was one of the earliest criticisms of Synthesis adaptationism. The most prominent paper of this genre was Stephen Jay Gould and Richard Lewontin's "The Spandrels of San Marco and the Panglossian Paradigm: A Critique of the Adaptationist Program." The spandrels paper identified the atomistic view of traits as the very first flaw of adaptationism: "An organism is atomized into 'traits' and these traits are explained as structures optimally designed by natural selection for their functions" (Gould and Lewontin 1979: 586).

This attribution may have been accurately applied to some adaptationists (perhaps to the sociobiologists who were Gould and Lewontin's primary target), but it was simply false of Mayr and Dobzhansky. Indeed, it wasn't even true of T. H. Morgan. We saw how Mayr argued against atomism in

1966 (Chapter 10, Section 10.6). Four years later (and nine years before the spandrels paper), Dobzhansky stated that "Talking about traits as though they were independent entities is responsible for much confusion in biological, and especially in evolutionary, thought" (Dobzhansky 1970: 65).

Dobzhansky and Mayr were opposed to atomism, and they acknowledged the interactive complexity of the relations between genes and traits. Nevertheless, this did not lead them to investigate the relevance of development to evolution, nor to question whether the Mendelian blind spots were hiding something important. Both authors do occasionally acknowledge the second Mendelian blind spot, which is intersterility as a barrier to the identification of genes between species. Their discussion of this issue shows a reason for their lack of concern about the blind spots and the Developmental Paradox.

Homologous genes are the simplest phenomenon that might be hidden by the blind spots. Intersterility blocks the discovery of homologous genes. Isn't the existence of homologous genes an important question? In 1966, Mayr considered the possibility that widely shared characteristics were caused by homologous genes. He rejected it.

> Much that has been learned about gene physiology makes it evident that the search for homologous genes is quite futile except in very close relatives (Dobzhansky 1955). If there is only one efficient solution for a certain functional demand, very different gene complexes will come up with the same solution, no matter how different the pathway by which it is achieved. (Mayr 1966: 609)

Two claims are involved in this statement. The first is that "gene physiology" makes the search for homologous genes futile. The second is that similarities should be presumed to have come about *not* by the action of homologous genes but by adaptive convergence. Convergent traits should be presumed to be the product of new genetic causes, assembled by selection, rather than by the old genes that had caused ancestral traits.

How does gene physiology imply the futility of the search for homologous genes? Do the facts about gene action imply the extreme modifiability of genes, so that they change too fast to be shared among distant relatives? No. Mayr documents his claim about gene physiology by referring to Dobzhansky (1955). Dobzhansky's discussion of gene homology makes two points. First, even within a single species, different genes can determine the same traits; several different mutations can produce the same mutant eye color in *Drosophila*. Second, we cannot operationally identify genes across species lines because of intersterility (except in a few cases of close crosses that yield sterile hybrids; see Dobzhansky 1955: 248–249). Thus, the only fact of "gene physiology" that makes the search for homologous genes futile is intersterility! This is the second Mendelian blind spot. Because we cannot test genetic homology by crossing, there is no way to tell whether the genes that determine a common trait are homologous or analogous. Notice that this fact about intersterility provides no grounds at all to doubt the *existence* of homologous genes – only to doubt their detectability. Nevertheless, Mayr follows his statement about futility with a statement about the power of natural selection to find alternative genetic pathways to any given adaptation. The relation between these two statements is problematic. On the more charitable reading, Mayr reports that the discovery of homologous genes is blocked by intersterility *but* that this fact doesn't matter anyhow, because natural selection can always achieve its results without homologous genes. On a less charitable reading, Mayr's statement about futility is misleading. The reference to the power of natural selection leads the reader to believe that the evidence for the power of natural selection is evidence *against* homologous genes; it is futile to search for homologous genes because we have evidence that they do not exist. In fact we had no such evidence. The evidence was hidden behind the Mendelian blind spot.

Dobzhansky does acknowledge some circumstantial evidence for homologous genes (e.g., in mammals a particular complex pattern of albinism arises independently in many species). However, Dobzhansky shares Mayr's intuition that (in the absence of evidence to the contrary) we should assume that similarities are due to adaptive convergence, not to common internal causes. He agrees with Mayr: If a trait (even an enzyme) is needed, evolution will find a way to produce it, and it will be able to find any number of different genetic means to do so.[60] Dobzhansky clearly recognizes that intersterility is a barrier to the discovery of homologous genes, but it doesn't worry him. Natural selection produces interspecies *diversity*; similarity must be a secondary effect of later convergent selection. Dobzhansky admits the bare possibility of the conservative retention of homologous genes, but he is not at all drawn to the idea. He certainly does not imagine that broad similarities across the animal phyla are to be attributed to the long-term sharing of genes.

This Mendelian blind spot was partially lifted in 1966, when the development of electrophoresis allowed the visible identification of distinct gene products by their molecular weights (Hubby and Lewontin 1966). In describing the comparative results of electrophoretic studies, Dobzhansky said that the technique "overcomes, at least to some extent, *the most serious limitation*

---

[60] Dobzhansky does acknowledge the "one gene, one enzyme" hypothesis developed by George Beadle in the 1940s, but he suggests that the principle applies only locally (Dobzhansky 1955: 249). This was much more reasonable in 1955 than it appears now. We now accept that enzymatic structures are "coded" in DNA sequences (Morange 1998: 27).

of the methodology of Mendelian genetics – the taxa compared need not be crossable and capable of giving fertile hybrids (Dobzhansky 1970: 363; emphasis added). To my knowledge, this is the first time that a Mendelian blind spot was admitted by a neo-Darwinian to be a "serious limitation," and the description was only used after the limitation had been eased. In 1955 it had been no more than a nuisance. Both Mayr and Dobzhansky were perfectly willing to predict that common characters among species were the result of adaptive convergence, not homologous genes, even though they both knew that the Mendelian blind spots prevented direct evidence for (or *against*) that prediction. They had faith in the power of natural selection to be able to produce any phenomenon that homologous genes could produce. This was sufficient reason to discount the significance of homologous genes.

Homologous genes were not at the time an important theoretical dispute. By the time the deep developmental genetic homologies such as *Pax-6* were discovered in the 1990s, the issue had been forgotten.[61] Nevertheless, the stance of Dobzhansky and Mayr is instructive. It illustrates an aspect of scientific commitment that extends well beyond the data immediately at hand. Advocates of a particular theory tend to have an optimistic view of how far their theory will extend, even in the absence of direct evidence. If homoplastic traits are traceable to homologous genes, then natural selection plays a relatively smaller role in the explanation of the homoplasies. (Perhaps it merely increased the frequency in the population of a recessive homologous gene that had existed at tiny frequencies, hidden by heterozygosity.) However, if homoplastic traits are independently sculpted in each species, using different genetic routes, then natural selection plays a far larger role in creating the observed patterns. The selectionist favors an explanation that postulates less underlying commonality and more selection over an explanation that favors more commonality and less selection. Even in the absence of direct evidence, advocates of selection are willing to predict that facts that are yet to be discovered will confirm the overriding importance of natural selection.[62]

This methodological tendency is shared with the structuralists and everyone else, of course. Everyone believes that their favorite theory will turn out to be more important than its rivals, and that facts not yet known will turn out

---

[61] The issue had been forgotten by all but the evo–devo authors, who chided Mayr for his claim that the search for homologous genes was futile (Gilbert et al. 1996: 365).

[62] My account of "explanatory force" (discussed in Section 11.6) could explain the tendency of Mayr and Dobzhansky to doubt the existence of homologous genes. Their commitment to natural selection leads them to anticipate that the conditions for the force of natural selective explanations are maximally satisfied. Because homologous genes would reduce the explanatory force of selective explanations, selectionists tend to doubt homologous genes even in the absence of direct evidence (Amundson 1989).

to be favorable to the importance of their theory. The case illustrates how that particular tendency plays out for neo-Darwinism: It produces an expectation of minimal underlying commonality. For homologous genes in the 1960s, the commonality under consideration was genetic, not developmental. That was soon to change. Developmental constraints became a serious challenge to neo-Darwinism. The anticommonality tendency that we see in Mayr and Dobzhansky's attitudes toward homologous genes springs up again in neo-Darwinian reactions to developmental constraints. This, however, was far from the only factor. Advocates of neo-Darwinism could also deploy the central conceptual dichotomies of evolutionary thought to prove the incoherence of structuralist evolutionary thought.

11.2   THE FOUR DICHOTOMIES DEFEND THE SYNTHESIS

The four dichotomies discussed at the end of the previous chapter were extremely useful in the neo-Darwinian arguments against the relevance of development to understanding evolution. They are so finely targeted to this use that one must carefully keep in mind the fact that the dichotomies were not invented for the purpose.

11.2.1   Maynard Smith: The Germ Line–Soma Critique

John Maynard Smith's *Evolution and the Theory of Games* had no particular relevance to the constraints debates except that it dealt with a novel application of optimization theory. However, it was published shortly after Gould and Lewontin's spandrels paper (1979), which was literate and stylish, an inspiration to structuralists, and a target of derision for Synthesis biologists. That paper soon became a required citation for anyone who wished to prove that they had taken the possibility of developmental constraints into account. Maynard Smith made it clear that he had done so in the Introduction to his book on games theory. Even though he believed that development was important, he did not believe that it was important *to the study of evolution*. He also believed the irrelevance could be easily proven:

> After the publication of Darwin's *Origin of Species*, but before the general acceptance of Weismann's views, problems of evolution and development were inextricably bound up with one another. One consequence of Weismann's concept of the separation of the germline and soma was to make it possible to understand genetics, and hence evolution, without understanding development. (Maynard Smith 1982: 6)

So Weismann proved the irrelevance of development to heredity, and thence evolution. In 1927, embryologist Frank Lillie had invoked the name of Weismann as the last great advocate of the unification of development and heredity (Chapter 7, Section 7.5). In 1927, Weismann was the emblem of the *integration* of development and heredity; in 1982, he was the hero who had divorced the two fields. Why the change? Two reasons. First, the modern dichotomous view of the organism as a combination of genotype and phenotype was well entrenched in the modern mind. Embryogenesis was no longer conceived as an aspect of an individual's nature, but merely that process by which one's nature (one's genotype) was "expressed" in one's body. Embryology itself was almost forgotten, and Weismann's mosaic embryological theory was totally forgotten. This enabled a replacement of the historical Weismann (who conceived of heredity as an aspect of development) with the modern pseudo-Weismann (whose germ line–soma distinction seemed to anticipate the genotype–phenotype distinction). References to Weismann were absent from formative Synthesis literature. Later in the century, Weismann was recalled to mind – he was the person who had refuted Lamarckism with his germ line–soma distinction. In the face of the new structuralist challenges, it seemed to Maynard Smith and others that Weismann had refuted not merely Lamarckism, but *any* relevance of development to evolution.

11.2.2 Hamburger and Wallace: The Typological and Germ Line–Soma Critiques

Viktor Hamburger was the only embryologist invited to the 1974 conferences that formed the basis of Mayr and Provine's *The Evolutionary Synthesis* (Mayr and Provine 1980). Aside from Waddington (who was not invited), he is a suitable choice. He was an extremely productive embryologist and had written about the history of his discipline. He had studied with Spemann, and his postdoctoral associates Rita Levi-Montalcini and Stanley Cohen had won the Nobel Prize for work begun with Hamburger. Hamburger had had nothing to do with the Evolutionary Synthesis, and he knew why. His article takes a moderate tone. He acknowledges that the absence of embryology from the Synthesis was primarily due to embryologists' own disinterest, and the reluctance of many of them to accept the division between heredity and development. Because of the inability of embryologists to understand gene action during development, "the embryologists of this generation were not ready to come to the aid of the architects of the new synthesis" (Hamburger 1980: 103). Hamburger here originated the claim that embryology had been black boxed by the Synthesis, but he did so in a gentle manner. "I would assert that it has always been a legitimate and sound research strategy to relegate to a 'black box,' at least temporarily, wide areas that although pertinent would distract from the main thrust" (Hamburger 1980: 99). His final section was entitled "The Missing Chapter," and it discussed the ideas of Schmalhausen and Waddington, with a short paragraph (six sentences) about genetic assimilation. Hamburger's article cast no blame about the absence of embryology from the early Synthesis, but his metaphors of the *black box* and *missing chapter* each imply that the Synthesis *now* should integrate development. Waddington's 1953 complaint about the absence of embryology had been ignored. Hamburger's mild 1980 suggestion for future research was not ignored, but rejected.

Bruce Wallace is a prominent population geneticist and former student of Dobzhansky. He was invited to contribute an article to an anthology about disciplinary integration in the sciences (Bechtel 1986). His article was entitled "Can embryologists contribute to an understanding of evolutionary mechanisms?" (Wallace 1986). The answer was no. The paper was specifically written to refute Hamburger's claim about the black-boxing of development. Development didn't need to be black boxed because it was irrelevant to evolution in the first place. In refuting Hamburger, Wallace deploys two of the dichotomies discussed in Chapter 10, Section 10.7. He follows Maynard Smith's use of the germ line–soma dichotomy, and he adds the population thinking–typological thinking dichotomy.

Wallace's typological critique of Hamburger is based on the fact that Hamburger does not describe the genetic variation that exists in a population either before or after genetic assimilation occurs. He reports Hamburger as believing that the only variation in either the preassimilation or postassimilation population is nonheritable and environmentally caused. "Missing from this account is the very heart of neo-Darwinism; the above account of genetic assimilation is an example of what Mayr refers to as 'typological thinking'" (Wallace 1986: 50). I read Hamburger slightly differently than Wallace, but Wallace is surely correct that Hamburger pays no attention to populational variation. It seems to me that Hamburger merely describes genetic assimilation from the standpoint not of the population, but of the individuals within the population that are among the adaptively successful variants. He does not *state* that the population is genetically invariant, but his discussion does not touch on its genetic variation. He discusses the members who exhibit the environmentally caused variation in the first place, and those of their descendents whose ontologies genetically assimilate toward the new phenotype. In doing so, he ignores populational variation and surely does miss "the heart of neo-Darwinism." But is *failing to discuss* populational effects sufficient to diagnose typological thinking? The accusation appears to be set on a hair-trigger. In fact, I do not deny that modern developmentalists engage in typological thinking: I will soon argue that they do. Nevertheless, Hamburger gave no positive indication of typological thinking. He merely showed a greater interest in the ontogenies of individuals than in populations. Embryologists do that.[63]

Wallace's second argument (of 1986) shows that embryology is irrelevant to evolution even if embryologists are *not* typologists, at least for those organisms that sequester their germ lines. He gives a sketch of the dynamics of evolution that exploits two recent notions: the genetic program (from Mayr 1961) and the developmental program. Evolution is a matter of the persistence of germ lines. Wallace gives a series of diagrams in which the germ line comes in from the left (from ancestors) and exits to the right (to descendents). The soma is extended on a sort of stalk above the germ line with one-way arrows upward, indicating the absence of somatic influence on the germ line. The soma interacts with the environment. Whether or not the germ line (together with its "programs") is passed on depends on the soma's interaction with the environment. Because the soma itself does not causally contribute to changes in the germline, development is irrelevant to evolution.

> The development of the individual . . . is governed by the *developmental program* which that individual has inherited from its parents. If the *program* is successful in producing an adult, reproducing individual, that *program* . . . is transmitted to offspring who make up the subsequent generation. If not, the responsible *program* stops. The relative proportions of various *genetic programs* among the incoming germlines and the continuing germlines need not be identical: the lowest level of evolution consists, then, of changes in the frequencies of *genetic programs* – or, stated more simply, of gene frequencies. (Wallace 1986: 158; emphasis added)

---

[63] Mayr himself had applied the typology critique to ontogeny as early as 1959: "Any author who uses findings from the ontogeny of an individual to prove one or another evolutionary theory proves thereby that he completely misunderstands the working of evolution. To extrapolate from the individual to the evolutionary 'type' and its fate is, of course, still another manifestation of typological thinking" (Mayr 1959b: 8). The context of this quotation, however, does not support the notion that Mayr is rejecting the relevance of *development* to evolution. The passage is preceded by a rejection of those who would infer from the "goal-striving" or "purposiveness" in an individual to similar goal-striving in a lineage. The quotation does indicate an insensitivity in Mayr to Waddington's hope that epigenesis might be evolutionarily important, as did Emerson's letter of that same year to the Centennial Committee. However, it does not really reflect – in 1959 at any rate – a conscious, principled, and deliberate claim that Waddington (e.g.) is a typological thinker. To my knowledge, such principled rejections of development didn't occur before 1980.

Wallace does not attribute this argument to Weismann, but germ-line sequestration is the entire grounds for the alleged irrelevance of development. He does not explain the difference between a developmental program and a genetic program; they seem to be two ways of viewing the same entity. The program (whether we call it genetic or developmental) that exists in the germ line is not affected by the processes in the soma. Wallace's reversion to gene frequencies instead of genetic programs at the end of the quotation is elaborated in the article. He claims that gene-selectionists (like himself) and individual selectionists (like Mayr) are merely looking at two sides of the same coin. However, embryologists are looking at a completely different coin, and one that is irrelevant to the evolutionary question. In order for an embryologist's research to be relevant to evolution, the embryologist must answer this question: "How does a developing organism alter the genetic program carried by its originating germ cells?" (Wallace 1986: 160). The true answer is this: It doesn't. Because it doesn't, embryology is irrelevant to evolution.

The simplicity and resilience of the pseudo-Weismannian argument is remarkable. On the one hand, Wallace surely couldn't have believed that the structuralists were so naïve as to assume somatic influences on the germ line. On the other hand, I know of no detailed defense against the argument of the pseudo-Weismann, and we've seen it both from Wallace and Maynard Smith. How can something so obvious and important to the critics of structuralism seem so beneath comment to its advocates?

So far we've seen uses of the typological–populational and germ line–soma dichotomies to expose the irrelevance of development to evolution. The other two dichotomies can easily be found in Mayr's writing.

11.2.3 Mayr: The Proximate–Ultimate and Genotype–Phenotype Critiques

The suggestion that it is the task of the Darwinians to explain development . . . makes it evident that Ho and Saunders [structuralist critics] are unaware of the important difference between proximate and ultimate causations. . . . [U]ltimate causations (largely natural selection) are those involved in the assembling of new genetic programmes, and proximate causations those that deal with the decoding of the genetic programme during ontogeny and subsequent life. (Mayr 1984: 1262)

Violation of the proximate–ultimate distinction became a theme in Mayr's rejection of the claims of the structuralist critics of the Synthesis. I do not intend to question Mayr's response to particular developmentalist critics. A great deal of Synthesis bashing was going on in the early 1980s, and some of it was specious. However, Mayr did not restrict the proximate–ultimate criticism to particular authors. He applied it very broadly:

> When one analyzes their criticisms one discovers that [structuralist critics] made no distinction between proximate and evolutionary causations. (Mayr 1992: 28)

In 1993 John Beatty presented a detailed history of Mayr's use of the proximate-ultimate distinction in a special session of the ISHPSSB honoring Mayr's ninetieth birthday. Beatty showed that the distinction had originally been useful in Mayr's research. It was later use to defend the importance of evolutionary biology, "to make the point that there was more to biology than the study of proximate causes" (Beatty 1994: 349). Beatty did not mention Mayr's later use of the distinction to critique developmental approaches to evolution, as illustrated in the preceding quotations. When Mayr rose to respond to Beatty's paper, his only comment prior to taking questions from the audience was to made sure that this most recent application was on the record:

> I must have read in the past two years four or five papers and one book on development and evolution.... And yet in all these papers and that book the two kinds of causations were hopelessly mixed up. (Mayr 1994b: 356)

The 1961 introduction of the proximate–ultimate distinction had made no reference to developmental causation, either as proximate or ultimate (Mayr 1961: 1502–1503). The distinction had been designed to protect the naturalist studies of populational phenomena not from structuralist opponents but from molecular ones. The proximate nature of development appears in the 1980s, to counter a new opponent. Mayr never gives a hint of how it would be possible to relate development to evolution *without* committing the proximate–ultimate fallacy, so it is hard to resist the conclusion that Mayr believes that the irrelevance of development to evolution follows directly from the distinction itself. Ontogenetic causes are proximate causes, and apparently for that reason alone development is irrelevant to evolution.

Recall that in 1966 Mayr had endorsed Waddington's concept of the epigenotype. Waddington had specifically introduced the notion as an intermediary between the genotype and phenotype. Wouldn't that intermediary position place development somewhere between proximate and ultimate, at least *partially* relevant to evolution? No. In the 1980s Mayr is still willing to talk about the epigenotype, but the epigenotype has lost its independent status and collapsed into the genotype. He says that structuralist critics fail to recognize the following:

> [T]he genotype . . . is nothing but the other side of the coin of what Waddington has called so perceptively the epigenotype. All of the directions, controls, and constraints of the developmental machinery are laid down in the blueprint of the DNA genotype as instructions or potentialities. The implication that evolution is a matter of development and not of the genotype ignores the inseparability of the genotype and its translated product, the epigenotype. (Mayr 1984: 1262)

Mayr accepts Waddington's words, but not his meaning. The genotype–phenotype distinction holds strong. The epigenotype is merely the "translated product" of the genotype. This interpretation is not obvious in the 1966 book, where development (the epigenotype) can be read as an independent stage of organization that integrates the effects of individual genes, and so itself evolves along with the genotype. That was surely Waddington's intention. However, in the 1980s (if not earlier), all of the integrative and form-building qualities of development are written back into the genotype itself. This is made easier by use of Mayr's own "genetic program" metaphor, which packs DNA full of instructions and controls. The result is a reinforcement of the genotype–phenotype dichotomy, with development once again rendered invisible. The observable aspects of development are mere proximate causes, and its constructive, integrative aspects are packed away into the genotype. This renders development irrelevant to evolution.

11.2.4   Refutation by Slogan?

I do not want to give the impression that the entire force of the Synthesis resistance to structuralist arguments is captured in these brief critiques. However, their easy availability does help us to understand the relation between development and Synthesis thinking. On the one hand, there is little evidence for a conscious opposition to developmental thinking during the formation of the Synthesis, even up until the 1970s. On the other hand, the four dichotomies were in place well before the protests of the 1970s, and they are easily turned against the relevance of development to evolution – at least to evolution as the neo-Darwinian sees it. These dichotomies, developed between 1915 and 1961, appear as logical truths. They are the basic conceptual dichotomies that every student of evolution is assumed to accept and understand. The fact that structuralist evolutionary thought can be so easily shown to be inconsistent with these basic dichotomies shows that structuralism and the Synthesis really do have deeply inconsistent views of the nature of evolution and how it is to be explained. I will attempt to articulate the depth of this inconsistency.

11.3 POPULATIONS, ONTOGENIES, AND ONTOLOGIES

Conflicting theories often involve conflicting ontologies. They disagree about *what exists* (ontology), the furniture of the world. The ontology of eighteenth-century preformationism included tiny preformed germs but excluded vital formative causes; epigenesis excluded the germs but included the causes. The difference between modern adaptationist Synthesis evolutionists and structuralist evolutionists is subtler, but it is equally distinct. After sixty years of the Synthesis, population thinking is today well established, and few structuralists deny the central role of natural selection. The question is whether any *other* causal, explanatory theory exists, related to development, that contributes significantly to the understanding of evolution. The four aforementioned dichotomies imply that it does not. Structuralists claim that it does.

Structuralists often recognize the barriers posed by these dichotomies. Waddington saw that the genotype–phenotype distinction gerrymandered development out of the picture for evolution. He proposed the epigenotype as an intermediary. The epigenotype was at first ignored; then it was rewritten by Mayr into a mere synonym for the genotype. A more recent proposal is to expand the proximate–ultimate distinction to recognize not only functional (proximate) and evolutionary (ultimate) but also distinctly developmental concepts.

> Functional biology = anatomy, physiology, cell biology, gene expression
> Developmental biology = δ | functional biology | $/\delta t$
> Evolutionary biology = δ | developmental biology | $/\delta t$
> (Gilbert et al. 1996: 362)

Here developmental biology is the study of changes in functional biology, and evolutionary biology is the study of changes in developmental biology. The Causal Completeness Principle is written into this formula: Evolution is *defined* as changes in developmental processes. Such a formulation assumes but does not prove the relevance of development to evolution. In the same sense, however, the exclusive doctrine of proximate-versus-ultimate causation (together with the labeling of ontogeny as "proximate") assumes but does not prove the *irrelevance* of development to evolution.

These two expanded structuralist formulas are both based on a commitment to regularities and causal laws at the ontological level of developmental, ontogenetic systems. Epigenetics *as a field of evolutionary study* is legitimate only if such generalizations exist. The problem is that such generalizations can be seen to conflict with the ontological commitments of neo-Darwinian theory. I believe that structuralists have a larger methodological problem than most currently recognize: They seem to be unaware that their favored explanatory methods *really are* in conflict with the core of population thinking. In certain ways, structuralists *really are* typological thinkers.

Typological thinking is not the methodological sin that is depicted by Mayr, Wallace, Bowler, and other Synthesis commentators. This was already indicated in Part I of this book. Explanatory typology is a perfectly legitimate approach to the problem of form, but it is in conflict with population thinking, and that conflict must be recognized and eventually dealt with. I suspect that most structuralists would deny that they are typological thinkers; they have internalized the Synthesis condemnation of typology, and they seldom recognize it in their own work.[64] I will now try to explain the ways in which adaptationist and structuralist explanations diverge, in order to reveal the sense in which structuralists really are typological thinkers. Neo-Darwinian adaptationists and structuralists both conceive of individual organisms as merely bit players in the evolutionary drama. Individual organisms do not evolve. Evolution takes place in an abstract entity, one that is related to individual organisms but exists at a higher and more abstract ontological level than individual organisms. The nature of this entity differs between the two doctrines. The differences between these two evolving entities are very striking.

Adaptationist: Individuals don't evolve. Populations do.
Structuralist: Individuals don't evolve. Ontogenies do.

11.4 ADAPTATIONIST ONTOLOGY: HOW THE FOCUS ON
DIVERSITY AFFECTS ONTOLOGY

Modern adaptationist evolutionary biology is a tremendously diverse field, which any brief account will oversimplify. Nevertheless, in order to show the contrasts between evo-devo and adaptationist theory, I must choose some exemplary version of modern population-based evolutionary thinking. One modern approach to population genetics implies an antihistorical view of

---

[64] A notable exception was Pere Alberch. During the late 1980s a group of his graduate students referred to themselves as "the typologists" (personal communication with Gerd Müller, February 10, 2004, and Anne Burke, February 19, 2004).

evolutionary causation. This brings the contrast with structuralism and evo-devo into vivid focus. Recall how Dobzhansky considered morphological studies to be merely historical. To get at the *causes* of evolution, one must study genes in populations. Hudson Reeve and Paul Sherman (1993) similarly distinguish between studies of "evolutionary history" and of "phenotype existence" in a paper that is useful for the clarity of its critique of structuralist though. Adaptation can be studied with either approach, that is, direct population genetic studies or comparative studies to detect past episodes of selection. However, the contemporary "phenotype existence" studies are regarded as primary because they are said to study the causes of evolutionary change. Because the causes of change are populational, they can be directly studied only in populations.

I happened upon an excellent illustration of the antihistorical focus of population genetics while researching a paper on the history of the concept of adaptation. Philosophers of biology have almost universally endorsed the definition of *adaptation* according to which a trait is labeled an adaptation only when it has a history of selection for providing the fitness benefit that it now provides to a species. The view has been so dominant as to be called "the received view" (Brandon 1990; see Amundson 1996 for discussion). Surprisingly to the philosophers, several population geneticists rejected the historical definition. They preferred a definition according to which an adaptation was simply a character state that increased an individual's fitness in a population over its alternative character states (Endler 1986; Endler and McLellan 1988; Reeve and Sherman 1993). Past history was irrelevant to whether a trait was an adaptation. The quasi-purposive aspects of the historical definition of adaptation (the trait exists in a population *because* it was selected to do what it does) attracted the philosophers, but they didn't attract the population geneticists. The population geneticists used the term *adaptation* to designate a trait's *causal relevance* in contemporary population dynamics. Contemporary causal activity has nothing to do with past history. Population genetics is forward-looking only: "Whatever is important about a trait's history is already recorded in the environmental context and the biological attributes of the organism" (Reeve and Sherman 1993: 9).[65]

This conception of the importance of contemporary causal factors in populations appears to be at the core of the adaptationist rejection of nonpopulational thinking as methodologically flawed (e.g., "typological"). The *causes*

---

[65] The historical definition of adaptation is not particularly structuralist, and many adaptationists endorse it. Indeed, it is customarily credited to George C. Williams (1966). However, its can easily be used to generate paradoxes for adaptationism (Gould and Vrba 1982).

or *mechanisms* of evolutionary change can only involve these contemporary factors. All causally significant talk about past history is merely a description of the consequences of such events in the past. The philosophers' historical concept of adaptation makes it (adaptation) a label that attaches to a character in virtue of the character's past history. However, such history is causally irrelevant to the character's current selective status, its relevance *now*. Reeve and Sherman's statement about "whatever is important" is intended to refer to whatever is *causally* important. Records of past selection have no causal power over the present.

This all sounds harmless and commonsensical. Surely it is correct that a trait's past history does not free it from contemporary environmental demands. Populations change through time depending not on the *previous histories* of the traits their various members have at any time, but only on the aspects of those traits that *at that time* affect their bearers' fitness. In fact, though, this principle provides a real challenge to structuralist thought. If evolutionary causation depends on only the traits that are causally active in population genetic processes, then most of the characteristics of interest to structuralists are irrelevant to evolutionary causation! Structuralists believe that particular developmental characters, shared within taxa, explain something about the evolution of those taxa, but the fact that these characters *are shared* is a fact about history, not a fact about any current population of organisms within that taxon. Even if it were a fact about a current population, *that population (typically) does not vary* with respect to the developmental characters at issue. Therefore selection can have no effect on it. Therefore evolutionary causation must pass it by.

Diversity is at the center of adaptationist explanations, both because diversity among species is the typical evolutionary result and because heritable variation within the population is a necessary condition for any selective change. Recall the Mendelian blind spots; it was impossible to genetically identify traits that did not vary in a population, or that varied between noninterbreeding species. These were among the reasons that embryologists rejected Mendelism. Recall also that *these same blind spots didn't matter* to natural selection, applied through population genetics. As long as a varying trait was heritable, all other facts about the trait's history were irrelevant, including its embryological origin. A population of flies with heritable variation in bristle number is open to selection for bristle number. However, the causation of nonvarying traits is of no theoretical consequence. Heredity is undefined for these traits, and the lack of definition is inconsequential. Nonvarying traits may have varied in the past, and they may vary in the future. Nevertheless, for contemporary population dynamics they are simply background conditions.

A fly must have a back before it can have bristles on it, but selection is "blind" to nonvarying aspects of its back. Nonvarying traits are the canvas on which the adaptive evolutionary picture is painted. Admittedly, this canvas changes as traits become fixed, or begin to vary – but the explanatory action always takes place in the zone of variation.

This aspect of neo-Darwinian ontology can be traced back to Darwin himself. As I argued in Chapter 4, Section 4.9, Darwin's approach had as its goal the explanation of change rather than the earlier morphological goal of the explanation of form. Attention is focused on phenotypic variation within the population and selection upon this variation. The outcome of selection is a change in the frequencies of heritable traits (or genes, the representatives of traits). The change occurs to a population. As soon as speciation occurs, the population is irretrievably split in two, and the fates of the now-distinct species are separate. The mechanism of natural selection does not act on a pair of species, or on a genus or a family. It acts independently within each population. This is what it means to say "populations evolve." Natural selection may have similar effects on two species that have similar population structures and environments, but it does not act on these two species *as a pair*, and it would not have similar effects on them merely because they had a common ancestor. Once speciation occurs, no mechanism exists to maintain the integrity of the lineage; even to speak of "the lineage" is to label populations by their histories, not by any characteristic relevant to their current or future evolution. No selective mechanism can act on all and only the descendents of one ancestral species; no law of nature can apply to such entities as supraspecific taxa. Taxonomy is mere record keeping, and it involves neither laws nor causal explanations.

11.5 STRUCTURALIST ONTOLOGY: COMMONALITY AND DEVELOPMENTAL TYPES

The situation is very different in developmental biology. Most of the processes studied are fixed not only in a species but in high taxa. The processes shared within each stem and branch of the hierarchical tree of life are of interest, from those most widespread (e.g., early cleavage patterns) to those that differ between closely related species. Much research is devoted to anatomical items that characterize high taxonomic ranks, such as the neural crest or the vertebrate limb. Traits that vary within a population have an embryogenesis as well, of course, as the Causal Completeness Principle implies. Until recently, however, there has been little interest in these traits. The theoretical interest is in understanding how an organism develops out of a zygote *at all*, how it gets its segments, and its gut, and its legs, and its back. There is less interest in such developmental details as the number of bristles on one particular organism's back. Intraspecific variation is not entirely ignored among embryologists. The oldest continuing embryological interests in intraspecific variation is in teratology, one of Geoffroy's specialties (Alberch 1989). Monsters, of course, are notoriously out of step with Darwinian evolution. Some developmental studies in the 1990s began to focus on intrapopulational variation, but, as we will see, they too have commonality rather than diversity as their theme.

The traditional structuralist research goal of explaining form can be seen in modern developmental biology; how does adult form emerge during embryogenesis and later development? It is *possible* to interpret this project as the study of the proximate mechanisms involved in each ontogenetic event, each individual "decoding of the genetic program." In this interpretation, developmental biology explains how this chick developed out of its zygote, and how this fly did it, and this snail, and this sponge, and so on. However, to interpret developmental biology as a study of proximate causes is to ignore the emphasis that structuralists place on the commonality of developmental processes. Developmental theorists study *the vertebrate limb*, not this particular chick's wing. The vertebrate limb has been a subject of structuralist study for over 150 years. Although nothing approaching a complete theory has been proposed, sketches have been made, and it is possible to imagine what a full theory would be like. A structural theory of the vertebrate limb (if a complete theory were available) would apply to *all* vertebrate limbs, with more specialized theories addressing the limbs of vertebrate taxa. Modern sketches of such a theory began in the 1980s (though they harken back to Owen 1849). They distinguish between "permitted" and "prohibited" morphologies, and they infer these morphologies from what is known about mechanisms of limb development across the vertebrate lineage, as well as from observed interspecies variation (Holder 1983; Shubin and Alberch 1986; Hinchliffe 1989). Recent evo–devo studies have identified the molecular correlates of the fields that had earlier been identified as inductive organizers. The theory of the vertebrate limb would not be a proximate theory about the building of any single limb, or about the building of the limbs of a particular species. It would be a theory about the processes of limb embryogenesis, and how common and divergent elements of these processes range across a large chunk of the evolutionary nexus. It would reveal how the nested sets of homologies of limb morphology reflect the interplay of conserved and divergent form-generating processes in the embryos of tetrapods. With the recent discoveries in developmental genetics, the theory of the vertebrate limb might even relate it to the limbs of nonvertebrates (Shubin, Tabin, and Carrol 1997).

Students of limb development would consider such a theory to be relevant to evolutionary biology. The developmental theory of the limb can be deployed in explanations of homoplasy.[66] When two related species evolve different limb morphologies, this happens as a result of specifiable modifications in the particular processes that they had previously shared. An understanding of the morphogenetic processes will allow prediction and explanation of certain evolutionarily interesting phenomena (examples to follow). Given the developmental mechanisms by which limbs are constructed, some evolutionary modifications are developmentally likely, some unlikely, and some impossible. If the structuralist limb theory is a good one, it would help us to see which are which. Most structuralist studies do not consider variation within a population, but even when they do, the variation is seen to reveal an underlying continuity.

An example can be seen in studies of intraspecific limb variation among newts and salamanders. One such study examined 452 newts from a single population for variation in the configuration of cartilage and bone elements in the digits, carpus, and tarsus (Shubin, Wake, and Crawford 1995). Possible variations include the loss, addition, or amalgamation of the seven carpal or nine tarsal elements. Of the many possible variations, only a few were observed in the population. Forty percent of the variants represented typical configurations in other species. The most interesting were bilaterally symmetrical patterns. Because they are present in both left and right limbs, these indicate a global developmental influence. Of the five bilateral variants observed, two represented atavisms, reconstituting inferred ancestral morphologies. The other three represented derived conditions in nested clades of other urodeles. The important point is that this restricted pattern of intrapopulational and interpopulational variation can be seen to follow from quite specific, empirically discovered mechanisms of limb development. "Underlying developmental influences on anatomical variation may exert their effect on cladistic topology because of the structural hierarchy of *the urodele limb*" (Shubin et al. 1995: 882; emphasis added).

This talk of the urodele limb is not just a way of referring to the limbs of urodeles. *The urodele limb* is an abstract theoretical construct, like the

---

[66] Notice the contrast with the adaptationist account of homoplasy discussed in Section 11.1. Such developmentalist alternatives to adaptive convergence go back at least to Vavilov's Law of Homologous Series in Variation (Vavilov 1922; Spurway 1949).

bauplan, that expresses shared patterns of development. Its nature is inferred from comparative morphology and experimental embryology. This kind of explanation may make no reference whatever to selective forces. Even in studies of intrapopulational variation, there is often no interest shown in either the heritability or the fitness effects of the variants. From a study similar to the aforementioned one, "the observed pattern of variation from a variety of clades is consistent with Shubin and Alberch's model of chondrogenic pattern formation. A functional explanation is at least not required to explain the bias in the variation pattern" (Rienesl and Wagner 1992: 318; referencing Shubin and Alberch 1986).

Let us return to Mayr's statement that development was a matter of prox-imate causation. Notice that theoretical concepts such as the *urodele limb* are seen as embodying the hierarchically structured developmental mecha-nisms available within a taxon. This is why development is not seen as merely proximate. Even though the ontogeny of each individual salamander involves proximate processes, the urodele limb is an abstract theoretical entity that is embedded in a theory that links evolution to ontogeny. Thus conceived, the urodele limb preexisted even the selective processes that produced the modified limb of a particular urodele species. From this perspective, devel-opment (or its set of possibilities, as expressed in the limb) is more ultimate even than natural selection, because selection can act only on the variation allowed by the limb! Recall the discussion of Darwin's claim that Conditions of Existence was a "higher law" than Unity of Type (Chapter 4, Section 4.7). Adaptationists see structure as a mere consequence of previous adaptations; structuralists see adaptation as merely making adjustments on preexisting structure. Function and structure, the chicken and the egg.

My intention in this section is to explicate the sense in which modern structuralists really are typologists. They talk about entities that I call *devel-opmental types*. These include items such as the urodele limb and the neural crest. These entities are identified by hierarchically structured developmen-tal similarities among individual organisms. They are roughly identifiable as homologs (although complications in concepts of homology are discussed separately). Developmental types will roughly correlate to phylogenetic clas-sifications, but they differ in two ways. First, developmental types sometimes refer only to particular body parts or aspects of form. (The exception is the bauplan, or body plan, that designates the overall structure of the body.) Phylogenetic classifications take all characters of a species into account, not only those that have received morphological or developmental analysis. For example, the type *vertebrate limb* includes only limb-related aspects of the organisms.[67] Second, developmental types (unlike clades) are conceived to be causally relevant to the understanding of evolutionary change. Phylogenetic classifications merely record the results of the past changes that are associated with speciation events.

I refer to items such as the urodele limb as types for a second reason. It is to align the modern tradition with its predecessors. Developmental types play the same theoretical roles in evo–devo that morphological or structural types played in the nineteenth century; modern biology merely has a more dynamic understanding of the developmental basis of the types. In fact, the contrasts between these structural theories and their adaptationist counterparts also show some similarities. Recall (from Chapter 3, Section 3.2.2) how Geoffroy was forced to admit that he was dealing with "idealism" and "abstract entities" when he inferred that bones within species in distinct phyla were nevertheless, in some sense, *identical*. His structuralism violated the early-nineteenth-century principle I called the *empirical accessibility of function*, which restricted direct observation to observations of function and labeled hypotheses of structural correspondence as speculative and idealistic. The pendulum of epistemological fashion has swung back and forth at least twice since that time, and the principle no longer holds (especially following the development of molecular developmental biology in the 1990s). However, modern structuralist concepts violate a corresponding principle of relevance within population genetic theory. For modern adaptationists such as Reeve and Sherman, a characteristic can be causally relevant to evolution only if can influence the differential fitness of its bearers within a population. Developmental types do not designate properties that can take part in population genetic processes. Evolutionary causation (to an adaptationist) involves only populational processes, and so developmental types are causally impotent. This point is clearly stated in Reeve and Sherman's critique of David Wake's structuralist analysis of the frequency of homoplastic digit reductions among plethodontid salamanders (Wake 1991). Wake bases his analysis on how urodele limbs develop; he does not consider the fitness of the morphological variants. For this reason his critics contend that he offers "at best a description, not an explanation, of the occurrence of four-toedness" (Reeve and Sherman 1993: 22). Wake's explanation is based on the causal mechanisms involved in limb generation, shared among urodeles. He means to account for constraints *on form*, not constraints on fitness (Wake 1996; Amundson 1994).

---

[67] The very existence of developmentally distinct characters such as limbs has important consequences in evo–devo (Wagner 2001a).

Nevertheless, in the absence of a test of the relative fitnesses of the four- and five-toed forms, this is no explanation at all for the adaptationist. The structuralist hypothesis has significance only if an adaptationist explanation can be tested and shown to be false.[68] Without a refutation of adaptation, the structuralist "explanation" can be no more than a description. The structuralist notion that developmental types affect evolution appears to imply that ancestral species are exerting control over the evolution of their descendants. Reeve and Sherman vividly express their adaptationist disdain for this notion:

> Ancestral species do not . . . mysteriously reach from the past to clutch the throats of their descendants. (Reeve and Sherman 1993: 19)

This is a real cognitive clash. Structuralists see developmental types merely as the representatives of shared developmental processes. They do not find anything mystical or mysterious in the concepts. Reeve and Sherman interpret such entities as superstitious bogeymen, because they do not "really" exist in current populations. Developmentalists have their eyes on the distribution of the developmental processes making up the *vertebrate limb*, and they regard these processes as exemplified or expressed in individual species. From this perspective, nothing mysterious or throat-clutching seems to follow from the persistence, under modification, of developmental processes in related species.

We must not be too quick, though, to dismiss the populational perspective of Reeve and Sherman. The question is not whether it is reasonable to categorize species together when they happen to share a common characteristic. Adaptationists have no objection to fitness-related categories such as *predator* and *grazer* or functional analogies such as *wing*. Generalizations are fine, when the properties being generalized relate to fitness and adaptation, but the development categories do not qualify. Reeve and Sherman treat the developmental types as mystical because the characteristics that unite them (unlike those that unite *predators*) are not defined in terms of current fitness. The developmental types (bauplans and vertebrate limbs) are defined by historical and developmental criteria, not by criteria associated with the dynamics of selection within current populations. Structuralists seem (to the adaptationists) to be saying that past history is causally affecting present-day populations *without explaining how this can be accomplished*! Ancestral

---

[68] Recall Bell's 1833 dismissal of the ear–jaw homologies. He said that the correspondences would be meaningful only if the species with fewer ear bones was proven to have poor hearing (Chapter 3, Section 3.3.3). To an adaptationist, structuralist views are meaningless unless they account for adaptive failures.

species (from which the shared processes were inherited) seem to be reaching out and clutching the throats of their descendants.

Several aspects of the evo–devo perspective on (what I call) developmental types are expressed in the following passage from Brian Hall.

> It is not that biologists disagree over whether animals possess basic body plans, for such plans exist, but the developmental and evolutionary significance of basic body plans that is in contention.... While concept is not mechanism, the *Bauplan* is no more metaphysical than are the designations blastula, neurula, tadpole and larva.... To search for the mechanisms of metamorphosis is not to deny the existence of tadpoles and larvae. Rather, it is to use those well-recognized ground plans as the starting point in a search for the mechanisms that produce them and the adults that form from them.... The need is not to regard the *Bauplan* as the idealized, unchangeable abstraction of Geoffroy, but to treat it as a fundamental, structural, phylogenetic organization that is constantly being maintained and preserved because of how ontogeny is structured. (Hall 1999a: 98–99)

Hall takes the existence of body plans (bauplans) as given, not under debate. Hall assumes that disagreements concern only the evolutionary significance of bauplans. The recognition of bauplans (or other developmental types associated with life stages, such as blastula) is important, but only as a step toward a study of the processes that produce and maintain them, and that they are causally involved in (such as metamorphosis). He acknowledges the evolutionary structuralist's dilemma in his comment on Geoffroy's "unchangeable abstraction": Developmental types are real *even though* they evolve. They have causal influence *and* they also can be modified during evolution. His final sentence is the most perplexing from a modern adaptationist's perspective. The bauplan, a "structural, phylogenetic organization," is "constantly being maintained and preserved because of how ontogeny is structured." If a single bauplan is held in common by species as widely divergent, and as reproductively isolated, as *all mammals* or *all insects*, what kind of bizarre cosmic force could "maintain and preserve" it? There is nothing in the natural world that can exert a force on all and only insects, or all and only mammals. This is very mystical sounding indeed. If, that is, you begin with a population–genetic perspective on causation.

Bauplans are taken very seriously within evo–devo. The consensus is that there are about thirty basic bauplans, and the consensus is fairly robust (Raff 1996). They are not regarded as highly inferential but as one of the most basic facts of evolution. One of the central questions within evo–devo is that of *The Origin of Animal Body Plans* (Arthur 1997). Developmental types associated with body parts (limbs) or life stages (tadpoles) have a similar status. There are disagreements about details, of course. Some authors are willing to posit developmental types where other authors only see coincidental similarities, but it is assumed by everyone that developmental types such as bauplans and vertebrate limbs are causally involved in the evolutionary process in three ways. First, such a type shows a real unity that calls for a specific causal explanation (i.e., they are not mere coincidences, or epiphenomena). Second, the observed unities are to be understood in terms of developmental processes (even though no *simple* association between ontogeny and adult form exists; the biogenetic law is false). Third, once these unities are understood at the developmental level, we will have a much richer understanding of other evolutionary phenomena.

Nevertheless, the tension with population genetics still remains. The unities associated with developmental types are, all of them, exhibited across reproductively isolated populations. How can they be involved in evolutionary causation? Developmental types are involved in two kinds of evolutionary causation, upward (as they are maintained and modified during phyletic time) and downward (as they influence the phenotypic variations that are made available to natural selection in individual species). The only evolutionary causation recognized by adaptationists is selection within a population. Such causation can have no "upward" effects on a bauplan, because its effects are limited to an individual species. Furthermore, the bauplan can have no "downward" effects on an individual population simply because the bauplan is not a biologically real entity. From the population genetic perspective, this story sounds very much like ancestral species' clutching the throats of their descendants!

I am truly impressed by the incommensurability of these two approaches. Evo–devo authors claim to be working on issues important to the understanding of evolution, but they don't follow the rules that would allow their explanations to fit within population–genetic assumptions regarding causation. Each of their developmental types applies to taxa above the species level, the species of which are reproductively isolated from one another. What could "maintain and preserve" such an entity, distributed as it is across reproductively isolated populations that are extremely diverse in their adaptive fit? A population thinker, focused on variation and selection within populations, and conceiving of natural selection as the only evolutionary cause or mechanism, quite reasonably finds the bauplan concept mystical. Even though structuralists insist that the developmental types are the results of evolutionary processes and are themselves subject to evolutionary modification, the fact remains that they range over reproductively isolated groups. From the adaptationist perspective, this just doesn't make sense. Even if a developmental type were conceived to govern the forms of reproductively isolated groups, the claim that such a type can *evolve* is problematic if not incoherent. What mechanism could account for the evolution of a supraspecific entity? It certainly could not evolve by ordinary population genetics. Adaptationists are completely unmoved by the structuralists' insistence that types evolve. They often claim, for example, that the mere fact that *vertebrates evolve* is sufficient to refute the existence of a vertebrate bauplan. Ghiselin reports on Gould's advocacy of a concept of constraint that that is said to be embodied in "the developmental and hereditary apparatus." The refutation of this notion is laughably simple. "That move clearly will not work, for, as has been obvious all along, the developmental and hereditary apparatus itself evolves" (Ghiselin 2002: 289).

I will call this the *problem of the nomological range* of the developmental type. Developmental types (bauplans, the urodele limb, the neural crest) are conceived to exert lawlike, causal influences over populations that have been reproductively isolated from each other, sometimes for hundreds of millions of years. Even if we set aside the adversarial exaggerations, and we recognize that evo–devo authors consider developmental types to evolve, to have phylogenetic origins and sometimes extinctions, the nature of developmental types still remains anomalous from the population–genetic viewpoint. It assumes some internal structure of ontogeny itself, shared by reproductively isolated species. This structure not only retains its identity and its causal causal influence through speciation events – the strength of the influence is sometimes even increased. Forelimbs and hindlimbs of vertebrates are thought to be *more homologous* now than earlier in phylogeny (Roth 1984).

Nothing in population genetics licenses the conception of such unifying causes. Neo-Darwinism is founded on a set of parameters that are formalized in the equations of population genetics; natural selection is the only directional evolutionary cause or mechanism. Speciation creates reproductive isolation and so prevents natural selection from acting as a unifying force among species. What about Hall's claim that this mysterious force is a consequence of "how ontogeny is structured"? Well, the MCTH explains how heredity works and how traits are caused. Even if the epigenotype of one population has a causal influence on the ontogeny of the organisms within that population, how can its causal influence spread across reproductively isolated genera, classes, and even phyla? Traits are caused by (or controlled by, or associated with) genes (transmission genes to be exact), and genes participate directly in population genetics. Nothing can have the nomological range that is alleged of the urodele limb. If this doesn't convince you of the impossibility of overarching developmental laws' exerting their effects on reproductively isolated species, consider the four aforementioned dichotomies. Each shows the irrelevance of development to evolution. This mysterious, unifying, developmental cause just doesn't make sense.

If population thinking is a recognition of population-level selection as an evolutionary mechanism, we might consider *exclusive population thinking* as the position that no generalizations or concepts that refer to higher taxonomic levels are relevant to the causal understanding of evolutionary change. From time to time in this book I have complained that neo-Darwinians fail to informatively define *essentialism* and *typology*. Perhaps no such definition is needed. Perhaps essentialism and typology are merely placeholders for evolutionary mechanisms that are *not* reducible to populational processes. Because developmental types are definable neither populationally nor adaptively, they are prohibited by exclusive population thinking. This would explain the incommensurability between neo-Darwinism and the concept of developmental types.

11.6 CONCEPTS OF HOMOLOGY

An illustration of the dramatic difference between the neo-Darwinian and structuralist programs is well illustrated in distinct concepts of homology. After Owen, homology remained a central theoretical concept in evolutionary morphology, and an important evidential concept (as proof of common descent) in Darwinian and neo-Darwinian theory. It has become a hot topic as the result of recent developments (Hall 1994, 1999b).[69] Several concepts of homology are in play, but this section focuses on the two that most clearly distinguish neo-Darwinian theory from evo–devo. One is the traditional "historical concept" first articulated by E. Ray Lankester (Lankester 1870). The other is what I call the *developmental concept* of homology, best articulated in a version called the *biological concept of homology* by Günter Wagner (Wagner 1989a; Wagner forthcoming). The historical concept is the traditional neo-Darwinian concept. On the developmental concept, associated with evo–devo, individual homologies themselves become developmental types as defined in the previous section. The discussion of homology serves as a stand-in for developmental types in general.

---

[69] These include the growth of phylogenetic systematics (cladism), the increasing use of molecular phylogenies, and the discoveries of "deep homologies" in developmental genes (such as *Pax-6* underlying the development of eyes in widely diverse species).

11.6.1   The Historical Concept of Homology

Lankester objected to the idealistic aspects of Owen's definition. He wanted to eliminate the vagueness of Owen's definition of homology as "the same organ in different animals." He defined homologs as characters in two species that "have a single representative in a common ancestors" (Lankester 1870).[70] Mayr's version of the definition is this: "A feature in two or more taxa is homologous when it is derived from the same (or a corresponding) feature of their common ancestor" (Mayr 1982: 45). Just as Darwin transformed the archetype into an ancestor (with no remainder), so Lankester transformed Owen's morphological concept of homology into the historical concept, with no remainder.

It is important to many neo-Darwinians that there really is no remainder in the historical concept of homology – that *common history and nothing else* constitutes homology. This stance is taken in opposition to structuralists (and to Owen himself) to whom homologies are reflections of underlying developmental causes. The underlying cause of a homology can influence the trajectory of evolution by acting as a developmental constraint (or alternatively as a window of opportunity).[71] For neo-Darwinians, homologies are mere residue of ancestry, not the expression of an active contemporary cause. This view is expressed by G. C. Williams with regard to the bauplan concept (which can be seen as a global homology of body plan). Williams proposes a null hypothesis for the existence of a real bauplan: An exhaustive list of the characters of descendent species would show a random distribution through time of the characters that change and those that do not. (If only 10 percent of ancestral characters remain after 100 million years, then only 1 percent will remain over 200 million years.) The hypothesis of a real bauplan would imply the relative permanence of the characters that are embedded in the

---

[70] Lankester actually introduced the term *homogeny* as a replacement for *homology*, but it never caught on.

[71] The negative image of constraint as a restriction is itself a by-product of the dominance of adaptationism. Structuralists often consider ontogenetic structures as productive, not constraining. From a conference in the midst of the constraint debates, Horn et al. stated that "every time that someone mentioned a 'constraint,' someone else reinterpreted it as an 'evolutionary opportunity' for a switch to a new mode of life, and a third person would bring up the subject of the complementary 'flexibility'" (Horn et al. 1982). Evo–devo author Wallace Arthur has recently argued that the open opportunities for evolutionary change within a developmental system should been seen as progressive, a form of "developmental drive" (Arthur 2001). Philosopher Denis Walsh has recently argued that developmental constraint ought to be regarded as causally responsible for the evolution of adaptedness within individuals, whereas selective explanations are causally responsible for the spread of such individuals within populations (Walsh 2003). The choice of progressive versus restraining–constraining metaphors is the rhetoric of science.

bauplan compared with those that are not. If characters decay in the stochastic pattern, the reality of the bauplan is refuted. Williams expects it to be refuted. No persistent causal factors underlie homologies; they are merely the ancestral characters that happened by coincidence to survive (Williams 1992: 88).[72] It is not common to find this degree of complexity in adaptationists' discussions of homology. The historical account of homology seems clear and and simple. It does all that is required of it: it summarizes the evidence of common ancestry. Discussions like Williams' occur only in the context of critiquing the developmental concept of homology (see for example Ghiselin 2002). It seems clear, even in the absence of frequent discussion, that to most adaptationists the historical concept of homology amounts to a concept of homology *as residue*. They reject the interpretation of homology as a reflection of underlying developmental commonalities.

11.6.2   The Developmental Concept of Homology

The developmental view of homology, in its simplest form, states that homologies are reflections of shared developmental processes. The sharing of these developmental processes constitutes that additional evolutionary cause or mechanism that separates evo–devo from neo-Darwinism. Unlike the historical concept, the developmental concept depicts Owen's special homologies and serial homologies as different aspects of the same thing, the shared developmental processes. As the Causal Completeness Principle stipulates, evolutionary changes *are* changes in developmental processes. The fact that homologies persist through evolutionary changes means that they reflect aspects of development that are less malleable than others. They can be seen as probes of the developmental process. In fact, they may be probes not only of those aspects of ancestral ontogenies that are handed down to descendents (like Owen's homologies of the vertebrate limb) but also of the evolution of development itself. If it is true that tetrapod forelimbs and hindlimbs became more homologous through evolutionary time, this may be explained as the co-option of the limb-generation processes between the two pairs of limbs (Roth 1984: 22).

The developmental view of homologies is faced with serious factual hurdles. Homology is not reducible to any simple aspect of development. The nineteenth century saw a series of debates about the relative importance of

---

[72] Similar views on homology are stated by Ghiselin, who attributes the belief in developmental causes of homologies to essentialism and Platonic idealism (Ghiselin 1997: 213). Homologies are mere namesakes; those who believe them to be more fundamental than analogies simply "get it backwards" (Ghiselin 2002).

adult comparative anatomy versus embryological precursors as indicators of true homology (Owen vs. Huxley; Gegenbaur vs. Haeckel). The irregular behavior of homologs was behind the breakdown of evolutionary morphology (recall the despairing quotation from Adam Sedgwick in Chapter 5, Section 5.6). It has exercised the greatest embryologists (Spemann 1915; De Beer 1971). Homologous characters do not always arise at corresponding times in ontogeny, or develop out of the same embryological precursors or even the same germ layers, or reflect expression of the same genes. Information from anatomy, embryology, and genetics can give inconsistent answers when one is trying to determine the homology of a character. Regenerated organs are clearly homologous to those originally developed in embryos, but they are constructed in a different manner, and often from different tissue sources. The germ-layer theory and the biogenetic law had been the foundations of evolutionary morphology. The failure of these views strongly affected the willingness of structuralist thinkers to engage in evolutionary argument. De Beer asserted the Causal Completeness Principle even in the heyday of the Synthesis, and he argued for the relevance of embryology to evolution (De Beer 1938, 1951). However, he was still arguing against the biogenetic law in the 1950s, as if it were a contemporary adversary (Churchill 1980). His important study of the complexity of homology may be read as an explanation of why he couldn't do more to bring embryology into the mainstream of evolutionary biology, despite his insistence that it should be there (De Beer 1971). Had homology behaved itself, evolutionary morphology would not have failed in the first place.

By the 1980s, inspired by the constraints debates, biologists were again exploring the developmentalist concept of homology. The early discussions overlooked some of the complexities that had so frustrated Spemann and de Beer; homologies were explained in terms of common genetic control (Roth 1984). The important point was that homologies were characterized by contemporary developmental causes, not mere history. Günter Wagner began to pursue his biological concept of homology (Wagner 1989b; following Roth 1988). I give special attention to Wagner's account for two reasons: He applies it to a wide range of issues within evo–devo, and he gives an insightful philosophical account of the ontology of his program.

According to Wagner, a developmental account of homology should account for three things: the conservation of homologs through evolutionary time, the individuality of the body parts identified as homologs (i.e., their distinctness from other body parts), and "uniqueness" of the origin of new homologs (Wagner 1989b: 1163). By uniqueness, Wagner refers to the rarity of origin of a character that can then be identified as a homolog in descendents;

it is this rarity that allows developmentally defined homologs to characterize monophyletic groups.[73] Wagner does not offer a simpler, more direct, or more reductionist definition of homology than the earlier workers. Instead he broadly characterizes the kind of epigenetic organization that can produce well-individuated characters that persist (under modification) through evolutionary time. These must involve networks or hierarchies of induction relations within the developing embryo (following the kinds of developmental integrating processes discussed in Alberch 1982 and in Sander 1983). Such networks can produce relatively persistent phenotypic results even through the kinds of genetic and embryological modifications that had refuted simpler developmental accounts of homology. Genes are crucial actors in these networks, but genes exert their effects indirectly, through the physical interactions (e.g., induction) of the cells in which they are expressed.

Wagner's concept of homology was intended not only to account for the homological "sameness" of body parts in different species, which was the goal of earlier definitions of homology. Wagner was also concerned with accounting for how bodies come to be individuated into parts in the first place. This second interest led him to edit an anthology on the *character concept* in evolution, in which character is taken to refer to a body part that is sufficiently individuated to be identified under a wide range of modifications (Wagner 2001a). The individuation of body parts (the creation of so-called characters) is an obvious precondition for homology in its traditional anatomical meaning. By making it a part of the theoretical definition of homology itself, Wagner is elevating the status of homology even within evo–devo. Homology is not just the recognition of correspondence but the explanation of correspondence, and in fact it is the explanation of why organisms evolve body parts (characters) that *can* correspond. An illustration of the new importance of homology can be seen in Wagner's interesting critique of the historical definition of homology.

Let us begin again with Owen's definition of homology as "same organ." Owen's only explication of sameness was by reference to the Vertebrate Archetype, which (idealistically interpreted) was objectionable to Darwinians. The historical definition replaced Owen's question-begging sameness with "feature derived from the same feature of their common ancestor." This was supposed to be an improvement in the specificity of sameness, because ancestors (unlike archetypes) are tangible entities. However, Wagner shows that the historical definition *does not solve the problem* of specificity of homology.

---

[73] The characterization of monophyletic groups is the definition of homology according to the *taxic concept* associated with phylogenetic systematics. It is a consequence of developmental uniqueness according to Wagner's concept.

The historical definition falsely asserts that the body parts of descendents are "derived from" the body parts of their ancestors! This is manifestly false for any theory of heredity since the eighteenth century. Legs are not derived from legs, and heads from heads. Body parts are epigenetically built anew in each generation (Wagner 1989b: 1159).[74] Because body parts are never derived from ancestral body parts, the historical definition fails to specify the grounds for identity in homology just as badly as Owen had. The materialist-looking reference to the body parts of real ancestors fails to pick out which body parts *correspond*. The attempt to specify ancestral body parts as those from which modern body parts are *derived* is preformationist nonsense.

One might think that this a mere quibble. Surely a neo-Darwinian could easily redefine historical homology to avoid the misstep of the preformationism about body parts – but it's not as easy as it seems to extract the preformationism.[75] How might one establish the identity of body parts between species without imagining body parts reproducing other body parts? Not by claiming that the three species' body parts are under the control of homologous genes. Long before we could identify genes across species lines, it was known that homologies were not necessarily caused by the same ones (De Beer 1971). The plausible option for a neo-Darwinian is merely to claim that one can recognize corresponding body parts when one sees them, and that certain parts that correspond between modern species *are explained* by the correspondence of each with their common ancestor. This simply begs the question of how correspondence is recognized.

If the neo-Darwinian avoids body-part preformationism and merely alleges that correspondences in body parts are recognizable, two questions remain:

1. Why do organisms have distinct and recognizable parts at all?
2. Why is it so easy to spot correspondences among them?

There is no pressing demand within neo-Darwinian theory to address these questions. Because homology is mere residue, its explanation has no great importance.

The interesting point is this: *Wagner intends to answer these questions*. The biological concept of homology is proposed to explain, in one fell swoop, why bodies are divided up into characters, and why these characters identifiably persist through long stretches of phylogenetic time. If such an explanation is successful, it will contribute to our knowledge of the causal basis of

---

[74] This shows how the conflict between neo-Darwinism and evo–devo regarding homology reproduces the preformationist versus epigeneticist controversy (Maienschein 1999).

[75] The taxic definition easily avoids preformationism, but that is not under consideration at the moment.

developmental types. It will explain the actual mechanics in ontogeny that lie behind the persistence of these types through the immense diversification involved in the evolutionary history of large groups such as vertebrates or insects. It will fill in the causal details behind Hall's description of the bauplan as "a fundamental, structural, phylogenetic organization that is constantly being maintained and preserved because of how ontogeny is structured."[76]

Wagner's account of homology connects with two more recent topics that have become areas of active research in evo–devo. One is the concept of modularity, which can be seen as a generalization of Wagner's notion of characters. Modularity is produced by the evolution of genetic and epigenetic controls over a part or aspect of an organism's body so that the part (the module) develops and evolves quasi-independently from the rest of the organism. This is thought to enhance the "evolvability" of the organism. Variation within a module need not disrupt the functioning (either the embryological development or the physiological function) of other modules; an individual module can "explore" its phenotype space without affecting the entire body. The other is the explanation of evolutionary innovation, a long-standing difficulty within neo-Darwinian thought (Robert 2002). Wagner's discussion of innovation includes the philosophical interpretation of the debates he is engaged in, and it is sensitive to what I have been calling the incommensurability of neo-Darwinian and evo–devo explanations. First, a bit of autobiographical background.

11.7 A PHILOSOPHICAL ONTOLOGY OF EVO–DEVO

I have for many years puzzled over the inconclusive nature of the debates between neo-Darwinians and structuralists. Even though natural selection and developmental constraints seem to take the form of opposing forces in these debates, the "force" metaphor was not helpful in analyzing the debates. The formulae of population genetics include parameters for selection, drift, and mutation, so an argument of selection versus drift was in principle decidable by empirical evidence. This is not so for selection versus developmental constraint: Development (and therefore constraint) is not a parameter in the population genetics formulae. I suggested, in 1989, that it would be more productive to abandon the attempt to estimate the objective forces of selection and constraint and to try instead to find ways of estimating the *epistemic* forces

---

[76] Hall is, of course, aware of these implications. My use of his quotation was illustrative only, not a suggestion that he has any larger problems than any other evo–devo practitioner.

of the respective explanations. Under what conditions is a selective explanation most significant or forceful, and under what conditions is a constraint explanation most forceful? This reflected my feeling (which persists in this book) that the difficulty of these problems is due not only to the complexity of nature but also to the logical shapes of our various theories about nature. The objective facts of population genetics and developmental genetics are both at work in every evolutionary phenomenon, but we have no way of measuring the force of one as compared with the force of the other. Instead, we should try to determine the circumstances in which the respective *explanations* have force; explanatory force.

One of the contemporary themes of evo–devo is the explanation of evolutionary innovation; this was a prominent topic in the symposium that heralded the creation of the Division of Evolutionary Developmental Biology in the Society for Integrative and Comparative Biology.[77] Günter Wagner later applied my notion of explanatory force to the issue of innovation. He chose two evolutionary explanations and evaluated the explanatory force of both selection and development for each. The two phenomena were (1) the evolutionary maintenance of stable 1:1 sex ratios in sexually reproducing species, and (2) the evolutionary innovation of eyespot patterns on butterfly wings (Keyes et al. 1999). Sex determination was a classic success of selective explanation, first proposed by R. A. Fisher. It applies to the large majority of sexually reproducing species, under a wide variety of ecological circumstances and ontogenetic mechanisms of sex determination. Butterfly eyespot patterns are a recent innovation with ecological significance in predator avoidance. Nevertheless, the evo–devo explanation given for their origin makes no reference to selection. The origin of eyespots is attributed to a modification of the developmental genetic system used in insects to construct the compartments of their wings. The original system (shared with *Drosophila*) uses many interacting genes to establish morphogenetic axes and boundaries. David Keyes and colleagues determined that small changes in two gene interactions (*hedgehog* and *engrailed*) had produced an "organizer" in the wing that produced a circular pattern in midwing (compare Spemann's organizer, Chapter 9, Section 9.2). Wagner points out that the selective explanation seems to have greater explanatory force in the sex ratio case, whereas the developmental explanation has more force in the eyespot case. Why should this be?

Wagner's explanation has to do with the *projectability* of the respective predicates of population genetics and developmental genetics in the two cases.

---

[77] The symposium was published as the November 2000 issue of *American Zoologist* (vol. 40, no. 5).

The sex ratio case operates perfectly well on the basis of transmission–genetic identifications of genes alone. The developmental basis of sex determination varies among organisms, but the "genes for" sex determination are subject to the same selective regime whatever their particular nature happens to be. However, transmission genetics is impotent in the eyespot explanation. The characteristics of *hedgehog* and *engrailed* that are involved in the explanation are indefinable in transmission–genetic terms. The relation of these genes to the phenotype is restructured during the developmental innovation itself: "Hence the transmission genetic properties of butterfly pigment patterns are not projectable; they change so radically that any attempt to explain them in the context of transmission genetics . . . is not very informative" (Wagner 2000: 97). For the innovation to be explained, a level of description must be chosen that remains invariant during the process being explained. Transmission genetics cannot be that level. Innovations in pattern formation that happen as a result of changes in the ontogenetic production of patterns will generally follow this pattern. When the developmental role of genes changes, then transmission–genetic definitions of those genes must change. For that reason, population–genetic explanations are blocked: They cannot pick out the "same gene" before and after the innovation. Put in another way, transmission genetics works on the assumption of the invariance of genes during the processes of reproduction and natural selection. When this invariance is violated (as it is during the change in developmental function that takes place in cases like the evolution of the eyespot pattern), transmission–genetic explanations lose their traction. Developmental genetics uses criteria of individuation for genes that are invariant during this kind of transformation. It can identify the so-called same genes even when the phenotypic effect of the genes has been modified. Innovations can be explained under evo–devo because developmentally conceived (and molecularly identified) genes are identifiable both before and after the innovation. In contrast, that kind of developmental genetic detail would make no difference in the sex ratio case. Sex is determined by different developmental causes in different species; natural selection affects only the frequency in each population, not the developmental properties, of the various genes involved in sex determination (and by hypothesis none of the genes will modify their developmental properties during selection). Therefore, each species' mode of sex determination is invariant during the course of selection, and the transmission genetic properties are sufficient to explain the maintenance of sex ratios without any reference to developmental mechanisms.

I find Wagner's analysis forceful in illustrating the contrast between the population–genetic and the developmental genetic, evo–devo styles of explanation. The theoretical differences among the various fields that study what is called "genetics" has been very great during the twentieth century (see Beurton, Falk, and Rheinberger 2000; especially Gilbert 2000). Neo-Darwinians may still insist that the developmental explanation is only descriptive, because selective processes have been ignored. Nevertheless, *something* was unquestionably explained by Keyes's research, and it had to do with the evolution of the eyespot.

Nevertheless, the example is deceptively simple. It disguises the conceptual gulf between the two views. The Keyesian explanation of eyespots is so convincing *only because* the developmental genes that underwent "reprogramming" could be identified by molecular means both in primitive and derived clades of butterflies. Many of the dramatic molecular homologies discovered in the 1990s were of the same nature (recall the list in Chapter 1, Section 1.2). We are now certain that evolution has proceeded by modifying development because we have the ability to identify the actual DNA sequences of the genes that operate in very early development. The problem is this: Naming the genes is not the same as explaining *how* development was modified to produce evolution. Even worse, naming the genes does not explain how the developmental types into which these genes are integrated were themselves "maintained and preserved" over evolutionary time. The real goal of evo–devo is to explain evolution *as* the modification of developmental processes, not merely to demonstrate *that* evolution proceeded by modifying development. Although genes are important aspects of the developmental processes, they are not the processes themselves. Furthermore, the conflicts between evo–devo and neo-Darwinian explanations are much more apparent when evo–devo advocates don't have their flashy new molecular homologies to fall back on.

Let us continue with Wagner's program. The biological concept of homology is intended to explain the origin of *characters*, the evolution of those developmentally modular body parts, like heads and limbs, that persist through immense periods of time and the radical restructuring of other aspects of the bodies of organisms. Although many of the molecular homologies are related to anatomical homologies, it is well known that anatomical homologs can arise ontogenetically by different embryological pathways. This means that the processes of ontogeny are integrated in such a way that a structurally similar phenotypic outcome can be maintained (and identified as homologous) even when the embryological processes that had originally built it are reshuffled. The discovery of molecular genetic identities, no matter how surprising and widespread, cannot explain this kind of integration. The integration must be understood as a process.

In the Introduction to his anthology on the character concept, Wagner discusses the philosophical notion of natural kinds, alluding to the work of philosophers from W.V.O. Quine to Boyd and Paul Griffiths, in a discussion similar to that in my Section 1.5.3 (Quine 1969; Boyd 1991; Griffiths 1997; Wagner 2001b). Natural kinds are categories that are discovered to be invariant, in some important way, under some set of processes that Wagner calls "reference processes." The standard example is chemistry and the discovery of chemical elements. As research proceeds, a set of phenomenal or operational laws is determined regarding the behavior of particular kinds during the reference processes (chemical composition and decomposition reactions for the elements, with the periodic table summarizing the relations). At this stage, researchers might propose what Boyd calls a "programmatic definition" of the kinds, specifying how the kinds are expressed in the phenomenal laws and the role they play in explanatory practices. The final step is a deeper mechanistic explanation of why the kinds exhibit the operational properties that they do. This will involve (for Boyd) an "explanatory definition" of the kinds. For those who interpret the developing theory in a realistic manner, either the programmatic or explanatory definitions might be said to designate essential properties of a kind.

Wagner discusses two ways in which the concept of natural kinds has recently been liberalized. One is the recognition that scientifically meaningful kinds need only be stable within a range of processes; they need not be fixed and eternal. (Chemical elements are still reasonably regarded as kinds after we discover the conditions under which they are impermanent; see Chapter 2, Section 2.2.) The other is that necessary and sufficient conditions (or correspondingly strict causal conditions) are no longer thought to be a requirement for a truly "natural" kind. A set of causal mechanisms that maintains homeostasis will suffice (e.g., Boyd's homeostatic property clusters; see Boyd 1991). Wagner discusses how the modern concept of species can be seen to fit this account of natural kinds; species function as the units of population–genetic evolution (explanatory role), their integrity is maintained by isolating mechanisms (homeostasis), and the homeostatic mechanism also allows one to understand the historical origin and fate of species.[78] Wagner wants to apply

---

[78] The treatment of species as natural kinds is inconsistent with the thesis of "Species as Individuals." This approach is defended in Boyd (1999). Although I have no strong feelings about the issue, it is interesting that the primary proponents of species as individuals, David Hull and Michael Ghiselin, were also central advocates of the Essentialism Story. Treating species as individuals can be seen as a way of maintaining the strict Popperian concept of essentialism according to which immutability and strict semantic definitions are required for natural kinds. Michael Ghiselin's *Metaphysics and the Origin of Species* assumes the correctness of this

the same analysis to the concept of character. Characters are units of anatomical evolution, just as species are units of population–genetic evolution. Like species, characters are real entities that are integrated by homeostatic mechanisms (in this case ontogenetic rather than populational mechanisms). They play an explanatory role in evolutionary theory. In fact, characters seem to be merely a special case of homology considered under Wagner's biological concept of homology. Both *characters* and *homologies* designate developmental types. The full account of the evolutionary and developmental origin and persistence of characters will account for their nature, in Hall's words, as "fundamental, structural, phylogenetic organization[s] . . . maintained and preserved because of how ontogeny is structured."

The aspect of Wagner's work that I find refreshing (as a philosopher, at any rate) is the forthrightness of his metaphysics. He is self-consciously arguing for a realistic interpretation of entities that have been marginalized within neo-Darwinian theory and historiography. This is, I believe, an essential step before evo–devo can be reconciled with neo-Darwinian theory. Wagner's comments about the species concept lead one to think that *character* and *species* are in some way parallel concepts within distinct research programs.

Adaptationist: Individuals don't evolve. Populations do. Species are effects of the evolution of populations.
Structuralist: Individuals don't evolve. Ontogenies do. Characters are effects of the evolution of ontogenies.

But I fear that the incommensurability persists. *Species* (perhaps more properly populations) are at the foundation of population genetics. *Characters* (or homologies – anyhow developmental types in one guise or another) are at the foundation of evo–devo. No one has quite explained how supraspecific developmental forces or entities can be maintained and preserved through evolutionary time, or exactly how neo-Darwinian theory was in error about the matter. Wagner's program of biological homology is aimed at explaining the evolution of modularity and the way in which (modular) characters contribute to evolutionary explanations. If this promise is fulfilled, it may solve the problem of the nomological range of developmental types. At the moment, though, the problem persists. Until our population-based evolutionary theory can be

---

metaphysical doctrine, and uses it to prove the incoherence of virtually all structuralist evolutionary thought (Ghiselin 1997). I must admit that the Popperian concept of essentialism seems to me like a dusty old relic of logical positivism.

reconciled with our homology-based evolutionary theory, we live without a true synthesis of evolutionary thought.

11.8 A NEWER SYNTHESIS?

Such are the conflicts between neo-Darwinism and evo–devo. The field of evo–devo is changing so fast that predictions about its future relations with neo-Darwinism are foolhardy. I end this study with a brief discussion of the current situation.

It is tempting to interpret evo–devo as an approach to macroevolution. Its practitioners study Unity of Type, and most of its central concepts deal with relationships among higher taxa, or properties (including developmental types) that are shared by these taxa. Indeed, I have identified the Extrapolation Principle as one of the barriers to structuralist approaches to evolution during the twentieth century. According to this principle, natural selection within populations is the sole mechanism of evolution, and macroevolutionary phenomena were the extrapolated results of these populational processes. Does this mean that structuralists were advocates of some alternative mechanism to natural selection as an engine of evolutionary change? Some were. Saltationist, orthogenetic, and several other macroevolutionary mechanisms had been proposed earlier, and they were soundly rejected by the Synthesis by means of the Extrapolation Principle. However, the principle did not *only* rule out such extravagant mechanisms as orthogenetics and saltational evolution. It reduced or eliminated the evolutionary significance of comparative studies of not only development but also morphology. Because transmission genetics embodied all of evolutionary causation, the outcomes of comparative studies could never be explanatory but only descriptive. Bodies themselves were important only as what would later come to be called "interactors" – as the items that were subject to selection. Their form and development were extraneous to evolutionary causation, because transmission genetics embodied all that was needed for that purpose.[79]

---

[79] This book has underreported the many comparative morphologists and paleontologists who advocated structuralism during the twentieth century. One strong protest against the dominance of the Synthesis, defending the significance of comparative morphology and paleontology, appeared in the very volume that celebrated the Darwin centennial (Olson 1960). The work of paleontologist Stephen Jay Gould is crucially important, especially *Ontogeny and Phylogeny* (Gould 1977). The most productive source of evo–devo thinkers during the second half of the century is arguably the Berkeley laboratory of comparative morphologist David Wake. Alan Love has documented the importance of morphology in keeping structuralist hopes alive during the twentieth century (Love 2003).

Is evo–devo likely to be associated with some alternative mechanism to natural selection to account for macroevolutionary phenomena? Developmental genetics has not run across any special mechanisms for macroevolution of the kind that Goldschmidt had proposed. However, some of the molecular discoveries seem almost to cry out for a saltationist explanation, such as the discoveries about the distribution of duplications of the *Hox* gene complex in different groups. These data, and a possible selectionist explanation of them, may be helpful in understanding the present state of evo–devo.

The number of *Hox* genes within a complex and the total number of *Hox* complexes have been studied in a wide variety of groups. A highly suggestive correlation can be seen between the number of complexes (and sometimes of genes within a complex) and the origins of major new groups. All invertebrates including basal ("primitive") deuterostomes such as echinoderms have one *Hox* complex. The hypothesized *Urbilateria* had seven *Hox* genes in its single complex. Duplication of the *Hox* complex had already occurred by the time of the transition between amphioxus and jawless fish; three complexes have been found in lampreys and four are present in all existing tetrapods. The genes within each complex have themselves duplicated and diversified to take on distinct developmental roles, with thirty-nine genes now in the four tetrapod complexes. The discovery of the developmental roles of these genes comprises much of the explosion of the 1990s (Carroll et al. 2001; Gilbert 2003a). The conclusion that the evolution of these higher degrees of complexity, and possibly the higher taxa themselves, arose *because of Hox duplication* is almost irresistible (Gerhardt and Kirschner 1997; Holland 1998). Similarly radical conclusions can be read from the developmental genetics of arthropods. The fossil record documents a history among arthropods of segment diversification of a kind that is mimicked by homeotic mutations (Budd 1999: 327). Could such mutations have been the literal origins of these groups? This approach, as tempting as it is, is radically at odds with population thinking. *Hox* genes are active at very early stages of development, when the basic outline of the body plan is being laid down. Modifications at that stage of development would (it would seem) produce extreme modifications in the adult, monstrosities, not the small variants that function nicely within ordinary populational selection. Are we to imagine that a single invertebrate duplicated its *Hox* complex and delivered its descendents to the base of the vertebrate tree? Like many earlier structuralist scenarios, this story ignores populational models of evolutionary change.

In one way, the modern discoveries of deep molecular homologies are very different from earlier structuralist views: They demand greater respect. They are themselves the outcomes of a *genetic* research program, not the (supposedly) discredited programs of the past.[80] To be sure, the more radical views, which associate *Hox* duplication or modification with the origin of higher taxa, meet considerable opposition. Graham Budd defends traditional populational explanations of morphological change against the structuralists who would explain it by large developmental genetic changes alone (Budd 1999), arguing that the *Hox* changes were caused by population processes after all. His argument contains an interesting irony.

Budd argues that even though the segment characteristics of arthropods are specified in the action of *Hox* genes, very early in the developmental cascade, the changes in segment identities that seem to indicate homeotic mutations had actually originated in downstream genes that had subtler morphological effects. The behavior of these downstream genes is far more amenable to populational and selective explanation. Budd proposes a notion of "homeotic takeover" to explain how population-genetic phenomena might, over time, result in modifications at a much deeper developmental level. One pattern seen in arthropod evolution is a change in segment type between a more posterior and a more anterior character. Each segment type is controlled by a specific *Hox* gene, so the change in appendage character is associated with change in *Hox* expression in the very early embryo. The more anterior segment is "taken over" by the *Hox* gene of a more posterior appendage. This appears to be a literal homeotic mutation. Budd proposes that it could have been originally produced in two stages, beginning with a phase of ordinary selection on the more posterior body part that caused it to resemble the more anterior body part. This stage occurs with no change in *Hox* expression. Once the posterior appendage is sufficiently similar to the anterior appendage, a shift in the homeotic domain might occur. Although homeotic mutations are dramatically "monstrous" events, this one was preceded by selection that produced phenotypic similarity between the old and the new domains. Thus, the homeotic shift was itself gradual in its phenotypic effects. The final result gives the appearance of a homeotic mutation at the origin of the taxon, but the actual change was gradual and selective throughout.

The irony in Budd's account is that "homeotic takeover" is a higher-level analog to Waddington's concept of genetic assimilation. Waddington had proposed that a character that had originally been environmentally induced might, through the modification of development, come under direct genetic control (see Chapter 9, Section 9.4.3). Budd proposes that a character originally controlled by downstream genes could be assimilated to *Hox* control

---

[80] Some believe that such "discredited" programs as experimental embryology deserve equal respect (Burian 1997; Gilbert 1996), but this is definitely a minority position.

(Budd 1999: 229). The irony is this: Waddington's advocacy of evolution by modification of development received very little approval from adaptationists during his lifetime, and genetic assimilation was reintepreted as ordinary selection on threshold effects. However, with the new molecular challenges from evo–devo, Waddington is suddenly recruited in defense of gradualistic adaptationism. Therefore, even if Budd is completely correct in his defense of adaptationism and populational mechanisms, it is a victory for evo–devo. Evolution is not merely the modification of phenotypes but the modification of ontogenies. As Waddington had argued all along, the epigenotype undergoes modification during evolution. The molecular discoveries have achieved that end: It is no longer possible to black box ontogenetic development. Unities of Type (revealed by developmental genetics) are staring us in the face.

I doubt that evo–devo will end up embodying an evolutionary mechanism that is a competitor to natural selection. The hot-blooded rhetoric of the 1980s is gone, and many of the leaders of the evo–devo movement soundly deny any revolutionary intentions (Hall 2000; Carroll 2001). A friendly unification with neo-Darwinism is foreseen, with little or no modification required of mainstream evolutionary theory. The macroevolutionary interests of evo–devo are not the kind that produce nonselective engines of evolutionary change. Nevertheless, there may be problems in forging a new synthesis. These come from the contrast in explanatory goals. Evo–devo may be seen to have a distinct evolutionary mechanism even though that mechanism does not compete directly with natural selection.

My attempt to account for the difference in explanatory goals (in this Chapter, and summarized in the paragraphs that follow) is influenced by comments made by two participants at a Dibner Seminar in the History of Biology entitled "From Embryology to Evo–Devo."[81] Historian Fred Churchill made the first comment while discussing the various structuralist theories of the late nineteenth century. The topic was Haeckel's notorious statement that phylogeny was the mechanical cause of ontogeny. Churchill said something to this effect: "The problem is in trying to understand what each of these theorists meant by 'mechanism.'" The second comment was from evo–devo researcher Rudy Raff, one of the earliest advocates of a developmental genetic form of evo–devo (Raff and Kaufmann 1983). In comparing his work with that of population biologists, Raff said that "they're interested in species; we're interested in bodies."

---

[81] The seminar was held during June 2001 at the Marine Biological Laboratory, Woods Hole, MA; it was directed by Jane Maienschein and Manfred Laubichler.

The phrase "mechanism of natural selection" falls trippingly off the tongue. This is merely an effect of familiarity; there is nothing machine-like about natural selection. There is no reason to require that every *evolutionary mechanism* must serve the same explanatory purpose as natural selection, must be an engine of evolutionary change. Structuralist programs have always had different explanatory goals from functionalist and Darwinian programs. They may, as well, refer to different things as "mechanisms." I therefore propose to interpret the Causal Completeness Principle as an assertion about an evolutionary mechanism. The central mechanism of adaptive evolution is natural selection within breeding populations. The central mechanism of the evolution of organic form is the modification of ontogenies. Neither mechanism answers the question that the other mechanism was designed to answer. Modification of ontogeny explains neither why the derived ontogeny occurred at the time it did nor why it was reproductively successful. Natural selection does not explain the organic form of either the primitive or the derived organism. Two mechanisms, two jobs.

Modification of ontogeny, offered as an evolutionary mechanism, may seem vague and trivial. If so, I encourage the reader to consider popular reactions at the time when the mechanism of natural selection was introduced. It was seen as trivial and tautological by many intelligent people who happened to be ignorant of the causal details that demonstrate the reality of natural selection in the real world. Natural selection, as it is now understood, is not merely the survival of the fittest organism. We now understand the heritability of traits; we can directly study their effects on fitness in real situations; we know the amount and the nature of heritable variation in natural populations; we recognize the affects of frequency-dependent selection; we understand isolating mechanisms and something about the genetics of speciation; and we know why phenotypically continuous variation does not imply hereditary blending of traits. Before we knew these things, natural selection was no more than a trivial (although very clever) conjecture.

We are only beginning to understand similar details about the *modification of ontogenies*. We know that the homeotic mutants discussed by Bateson and Goldschmidt were not evolutionarily irrelevant monsters, but indicators of developmentally deep regulatory genes. We recognize the difference between structural and regulatory genes, and we know that the dramatic effects of the "genes for" homeotic mutations occur because they are very early acting regulatory genes. The structural–regulatory distinction was unknown to the Synthesis architects. It was the first step toward the resolution of the Developmental Paradox, which began with Jacob and Monod's operon model of gene action (Jacob and Monod 1961; Morange 1998: 157). We also know that regulatory genes can *only* solve the Developmental Paradox because the action of each such gene in each cell is in part controlled by the cellular and embryonic environment in which these genes and cells find themselves.[82] Development is not merely a matter of genes building bodies, or even of genes influencing other genes to build bodies. It is a matter of the embryonic environment's influencing of some genes to influence other genes. These in turn modify the embryonic environment in an ongoing cycle, the ongoing result of which constitutes the organism. We recognize that evolution involves modifications in this ongoing cycle, modifications in the interactions within the developing body. We no longer identify genes by unique phenotypic effects (except, quaintly and anachronistically, by the effects that led to their first discovery – a mutation in *tinman* causes the heart to fail to develop). Instead we identify them by their molecular sequence and their performance(s) in ontogeny. We recognize that development is the product not of genes alone, and not only of the interactions between regulatory and structural genes, but also of the physical interactions of body parts within the embryo (that constitute the inductive processes studied by Spemann and others) and even of the interactions between the developing organism and its external environment (van der Weele 1999; Gilbert 2001).

Modification of ontogenies, like *natural selection*, sounds trivial only to those who do not understand the causal details that it summarizes.

Adaptationist: Individuals don't evolve. Populations do. Populations evolve by natural selection.
Structuralist: Individuals don't evolve. Ontogenies do. Ontogenies evolve by modifications of ontogeny.

Are these two programs compatible? Can we simultaneously think of populations evolving and ontogenies evolving?

My purpose in writing this book has been to provide a historical narrative that comes out right for evo–devo. I think I have done that, but at the very end, I see a problem. The Synthesis criticisms of structuralist theories have been

---

[82] Jacob and Monod claimed to have discovered "genetic regulatory mechanisms." This expression can be misread to imply that the network was wholly constituted of genes. A clearer expression would have been "mechanisms of genetic regulation" (Keller 2000a: 79). In their model, the action of regulatory genes was sensitive to environmental influences; the original model explained how bacterial gene action is modified in the presence or absence of certain nutrients. Although it is usually seen as a success of molecular biology, Jacob and Monod's model can also be seen as a vindication of the views of the embryologists who had opposed Mendelian genetics because of its refusal to address the Developmental Paradox; see Chapter 9, Section 9.3, and Gilbert (1996).

traced to their roots (in population genetics, the Extrapolation Principle, and the four dichotomies). However, these criticisms have not been refuted. They show that structuralist thought – in particular that involving developmental types – really is inconsistent with a certain version of population thinking. This version, *exclusive population thinking*, is a genuine problem for any future synthesis of evo–devo with neo-Darwinism. The problem remains even if natural selection is accepted as the only directional cause of evolutionary change.

Let us suppose that exclusive population thinking is generally accepted by neo-Darwinians. In this view, adaptive radiation is the way of evolution. Once speciation occurs, *no causal force* can unify distinct populations. Each of the four aforementioned dichotomies reinforces this point. Developmental types violate it. As long as development is conceived as a unified process that is shared among reproductively isolated groups, it is irrelevant to selection within populations. Therefore it is irrelevant to evolution. As long as evo–devo involves developmental types, it is perniciously typological. From this perspective, the only way for evo–devo to form a synthesis with neo-Darwinism is for evo–devo to abandon its fascination with developmental types. This means to treat homology as mere residue, to stop talking about entities such as the neural crest and the urodele limb (let alone the vertebrate bauplan), and to relinquish entirely the view that ontogeny is a thing that can be shared.

This is a very bleak view of the possibility of a future synthesis. Moderate evo–devo advocates are far more optimistic. They believe, for example, that population–genetic and developmental–genetic evolutionary explanations merely pertain to "different levels" of the evolutionary discussion. I think that the problem is larger than this, and that talk of different levels merely disguises the problem. Developmental types and exclusive population thinking are incompatible. One or the other (or both) must go before a new synthesis is possible. Here are two possibilities:

1. A way may be found to weaken the exclusivity of population thinking and so allow the use of developmental types. Wagner intends to establish an account of the homeostatic mechanisms within ontogeny that maintain the integrity of characters. This may sufficiently flesh out Hall's notion of a "structural, phylogenetic organization that is constantly being maintained and preserved because of how ontogeny is structured" that population geneticists will retract their opposition to the concept. Population thinking would be softened to allow a certain kind of typology.
2. A nominalist redefinition of developmental types may be devised, so that they are no longer conceived to range over reproductively isolated groups.

256

Exclusive population thinking would be retained and Wagner's realism about characters abandoned. Typology would turn out to be unnecessary after all.

I favor the first alternative, but that may merely be my structuralist prejudice. I suspect that the neo-Darwinian arguments are fallacious, but I cannot pinpoint the fallacy. A true defense of evo–devo may require the refutation of the dichotomies themselves.[83] But if the genotype–phenotype distinction is abandoned, what would become of population genetics?

Notwithstanding my methodological worries, the moderates are probably correct. History has a marvelous way of making philosophical and methodological difficulties disappear (poof!) in the face of scientific successes. Newton's gravitational force was action at a distance, which was metaphysically impossible. Within twenty years the metaphysical possibilities had changed. If both evo–devo and population genetics continue to be successful, a way will somehow be found to see them as consistent.

Structuralists and functionalists have been at odds since Geoffroy and Cuvier. A final resolution of the conflict of structure and function would be a truly momentous achievement.

[83] James Griesemer and others have begun to challenge the genotype–phenotype distinction (Griesemer 2000; van Speybroeck 2002). Heredity is said to be intertwined with development. Such notions have not been prominent since the 1920s, but as we have seen (Chapter 7), they were universal during the nineteenth century.

References

Agassiz, Louis (1962). *Essay on classification.* Reprint of 1859 edition. Edited by Edward Lurie. Cambridge, MA: Harvard University Press.

Alberch, Pere (1982). "Developmental constraints in evolutionary processes." In *Evolution and development.* Edited by John Tyler Bonner. New York: Springer-Verlag, pp. 313–332.

Alberch, Pere (1989). "The logic of monsters: Evidence for internal constraint in development and evolution," *Geobios, mémoire spécial no. 12*: 21–57.

Allen, Garland E. (1966). "Thomas Hunt Morgan and the problem of sex determination, 1903–1910," *Proceedings of the American Philosophical Society* 110: 53.

Allen, Garland E. (1978a). *Life sciences in the twentieth century.* Cambridge, England: Cambridge University Press.

Allen, Garland E. (1978b). *Thomas Hunt Morgan: The man and his science.* Princeton, NJ: Princeton University Press.

Allen, Garland E. (1985). "T. H. Morgan and the split between embryology and genetics, 1910–35." In *A history of embryology.* Edited by T. J. Horder, J. A. Witkowski, and C. C. Wylie. Cambridge, England: Cambridge University Press, pp. 113–146.

Amundson, Ron (1983). "E. C. Tolman and the intervening variable: A study in the epistemological history of psychology," *Philosophy of Science* 50: 268–282.

Amundson, Ron (1985). "Psychology and epistemology: The place versus response controversy," *Cognition* 20: 127–153.

Amundson, Ron (1986). "The unknown epistemology of E. C. Tolman," *British Journal of Psychology* 77: 525–531.

Amundson, Ron (1989). "The trials and tribulations of selectionist explanations." In *Issues in evolutionary epistemology.* Edited by K. Hahlweg and C. A. Hooker. New York: State University of New York Press, pp. 413–432.

Amundson, Ron (1994). "Two concepts of constraint: Adaptationism and the challenge from developmental biology," *Philosophy of Science* 61: 556–578.

Amundson, Ron (1996). "Historical development of the concept of adaptation." In *Adaptation.* Edited by Michael Rose and George V. Lauder. New York: Academic Press, pp. 11–53.

Amundson, Ron (2000). "Embryology and evolution 1920–1960: Worlds apart?," *History and Philosophy of the Life Sciences* 22: 335–352.

References

Amundson, Ron (2003). "Phylogenetic reconstruction then and now," *Biology and Philosophy* 17: 679–694.

Amundson, Ron (forthcoming). "Functions of myth: Essentialism as the foe of twentieth-century evolution theory," *Journal of the History of Biology*.

Appel, Toby A. (1987). *The Cuvier–Geoffroy debate: French biology in the decades before Darwin*. New York: Oxford University Press.

Aquinas, Thomas (1952). *Summa theologica*. Chicago: Encyclopedia Brittanica.

Arthur, Wallace (1997). *The origin of animal body plans: A study in evolutionary developmental biology*. Cambridge, England: Cambridge University Press.

Arthur, Wallace (2001). "Developmental drive: An important determinant of the direction of phenotypic evolution," *Evolution and Development* 3: 271–278.

Asma, Stephen T. (1996). *Following form and function: A philosophical archaelogy of life science*. Evanston, IL: Northwestern University Press.

Atran, Scott (1990). *Cognitive foundations of natural history*. Cambridge, England: Cambridge University Press.

Bacon, Francis (1960). *The new organon and related writings*. New York: Bobbs-Merrill.

Barrett, Paul H., Peter J. Gautrey, Sandra Herbert, David Kohn, and Sydney Smith (1987). *Charles Darwin's notebooks, 1836–1844*. Ithaca, NY: Cornell University Press.

Barry, Martin (1837a). "On the unity of structure in the animal kingdom," *The Edinburgh New Philosophical Journal* 22 (January): 116–141.

Barry, Martin (1837b). "Further observations on the unity of structure," *The Edinburgh New Philosophical Journal* 22 (April): 345–364.

Bateson, William (1922). "Evolutionary faith and modern doubts," *Science* 55: 55–61.

Beatty, John (1985). "Speaking of species: Darwin's strategy," In *The Darwinian heritage*. Editing by David Kohn. Princeton, NJ: Princeton University Press, pp. 265–281.

Beatty, John (1986). "The Synthesis and the synthetic theory." In *Integrating scientific disciplines*. Edited by William Bechtel. Dordrecht: Martinus Nijhoff, pp. 125–135.

Beatty, John (1994). "The proximate/ultimate distinction in the multiple careers of Ernst Mayr," *Biology and Philosophy* 9: 333–356.

Bechtel, William (1986). *Integrating scientific disciplines*. Dordrecht: Martinus Nijhoff.

Bell, Sir Charles (1833). *The hand: Its mechanism and vital endowments as evincing design*. London: William Pickering.

Bentham, Jeremy (1969). *A Bentham reader*. New York: Pegasus.

Beurton, Peter, Raphael Falk, and Hans-Jörg Rheinberger (2000). *The concept of the gene in development and evolution*. Cambridge, England: Cambridge University Press.

Bolker, Jessica A. (2001). "Evolutionary developmental biology: Developmental and genetic mechanisms of evolutionary change." In *Encyclopedia of the life sciences*. New York: Macmillan. *www.els.net* [doi: 10.1038/npg.els.0001517]

Bowler, Peter J. (1977). "Darwinism and the argument from design: Suggestions for a reevaluation," *Journal of the History of Biology* 10: 29–43.

Bowler, Peter J. (1983). *The eclipse of Darwinism*. Baltimore: Johns Hopkins University Press.

Bowler, Peter J. (1984). *Evolution: The history of an idea*. Berkeley, CA: University of California Press.

Bowler, Peter J. (1988). *The non-Darwinian revolution*. Baltimore: Johns Hopkins University Press.

References

Bowler, Peter J. (1996). *Life's splendid drama.* Chicago: University of Chicago Press.

Bowler, Peter J. (1999). "Evolution: History." In *Encyclopedia of the life sciences.* New York: Macmillan. www.els.net [doi: 10.1038/npg.els.0001517]

Boyd, Richard (1991). "Realism, anti-foundationalism and the enthusiasm for natural kinds," *Philosophical Studies* 61: 127–148.

Boyd, Richard (1999). "Homeostasis, species, and higher taxa." In *Species: new interdisciplinary essays.* Edited by Robert A. Wilson. Cambridge, MA: MIT Press, pp. 141–186.

Brandon, Robert N. (1990). *Adaptation and environment.* Princeton, NJ: Princeton University Press.

Bromberger, Sylvan (1966). "Why-questions." In *Mind and cosmos.* Edited by R. G. Colodny. Pittsburgh, PA: Pittburgh University Press, pp. 86–108.

Brook, Andrew (1994). *Kant and the mind.* Cambridge, England: Cambridge University Press.

Brower, Andrew V. Z. (2000). "Homology and the inference of systematic relationships: Some historical and philosophical perspectives." In *Homology and systematics.* Edited by R. Scotland and R. T. Pennington. London: Taylor & Francis, pp. 10–21.

Brush, Stephen G. (2002). "How theories become knowledge: Morgan's chromosome theory of heredity in America and Britain," *Journal of the History of Biology* 35: 471–535.

Buckland, William (1836). *Geology and mineralogy considered with reference to natural theology,* 2 vols. London: William Pickering.

Budd, Graham (1999). "Does evolution in body patterning genes drive morphological change – or vice versa?," *Bioessays* 21: 326–332.

Burian, Richard M. (1988). "Challenges to the Evolutionary Synthesis," *Evolutionary Biology* 23: 247–269.

Burian, Richard M. (1997). "On conflicts between developmental and genetic viewpoints – and their attempted resolution in molecular biology." In *Structures and norms in science.* Edited by M. L. Dalla Chiera. Dordrecht: Kluwer, pp. 243–264.

Burian, Richard M. (2005). "Lillie's Paradox – or, some hazards of cellular geography." In *The epistemology of development, evolution, and genetics.* Edited by Richard M. Burian. Cambridge, England: Cambridge University Press, pp. 183–209.

Burian, Richard M., Jean Gayon, and Doris T. Zallen (1991). "Boris Ephrussi and the synthesis of genetics and embryology." In *A conceptual history of modern embryology.* Edited by Scott F. Gilbert. New York: Plenum Press, pp. 207–227.

Burkhardt, Frederick, and Sydney Smith (1986). *Correspondence of Charles Darwin.* Vol. 2. Cambridge, England: Cambridge University Press.

Buss, Leo (1987). *The evolution of individuality.* Princeton, NJ: Princeton University Press.

Cain, Arthur J. (1958). "Logic and memory in Linnaeus's system of taxonomy," *Proceedings of the Linnean Society of London* 169: 144–163.

Camardi, Giovanni (2001). "Richard Owen, morphology, and evolution," *Journal of the History of Biology* 34: 481–515.

Cannon, S. F. (1978). *Science in culture: The early Victorian period.* Vol. 2. New York: Dawson and Science History.

Carpenter, William B. (1838). "Review of Whewell, 'A history of the inductive sciences,'" *British and Foreign Medical Review* 5: 319–342.

261

References

Carpenter, William B. (1889). *Nature and man: Essays scientific and philosophical.* New York: D. Appleton.

Carroll, Sean B. (2000). "Homeotic genes and the evolution of arthropods and chordates." In *Shaking the tree: Readings from nature in the history of life.* Edited by Henry Gee. Chicago: University of Chicago Press, pp. 69–88.

Carroll, Sean B. (2001). "The big picture," *Nature* 409: 669.

Carroll, Sean B., Jennifer K. Grenier, and Scott D. Weatherbee (2001). *From DNA to diversity.* Malden, MA: Blackwell Science.

Chambers, Robert (1844). *Vestiges of the natural history of creation.* London: Churchill.

Chung, Carl (2003). "On the origin of the typological/population distinction in Ernst Mayr's changing views of species, 1942–1959," *Studies in the History and Philosophy of Biological and Biomedical Sciences* 34C: 277–296.

Churchill, Frederick B. (1974). "William Johannsen and the genotype concept," *Journal of the History of Biology* 7: 5–30.

Churchill, Frederick B. (1980). "The modern evolutionary synthesis and the biogenetic law." In *The evolutionary synthesis.* Edited by Ernst Mayr and William Provine. Cambridge, MA: Harvard University Press, pp. 112–122.

Coleman, William (1964). *Georges Cuvier, zoologist.* Cambridge, MA: Harvard University Press.

Coleman, William (1976). "Morphology between type concept and descent theory," *Journal of the History of Medicine* 31: 149–175.

Coleman, William (1980). "Morphology in the evolutionary synthesis." In *The evolutionary synthesis.* Edited by Ernst Mayr and William Provine. Cambridge, MA: Harvard University Press, pp. 174–180.

Conklin, Edwin G. (1908). "The mechanism of heredity," *Science* 27: 89–99.

Crombie, A. C. (1994). *Styles of scientific thinking in the European tradition.* 4 vols. London: Duckworth.

Dana, James Dwight (1857). "Thoughts on species," *Annals and Magazine of Natural History* 20: 485–497.

Darwin, Charles (1859). *On the origin of species.* London: John Murray.

Darwin, Charles (1909). *The foundations of the origin of species: Two essays written in 1842 and 1844 by Charles Darwin.* Cambridge, England: Cambridge University Press.

Davis, D. Dwight (1949). "Comparative anatomy and the evolution of vertebrates." In *Genetics, paleontology and evolution.* Edited by Glenn L. Jepsen, George Gaylord Simpson, and Ernst Mayr. Princeton, NJ: Princeton University Press, pp. 64–87.

De Beer, Gavin R. (1938). "Embryology and evolution." In *Evolution: essays on aspects of evolutionary biology presented to Professor E. S. Goodrich on his seventieth birthday.* Edited by Gavin R. De Beer. Oxford, England: Clarendon Press, pp. 57–78.

De Beer, Gavin R. (1951). *Embryos and ancestors.* Oxford, England: Clarendon Press.

De Beer, Gavin R. (1971). *Homology, an unsolved problem.* Oxford, England: Oxford University Press.

De Robertis, Eddy M., and Yoshiki Sasai (1996). "A common plan for dorsoventral patterning in Bilateria," *Nature* 380: 37–40.

De Robertis, Eddy M., and Yoshiki Sasai (2000). "A common plan for dorsoventral patterning in Bilateria." In *Shaking the tree: Readings from nature in the history of life.* Edited by Henry Gee. Chicago: University of Chicago Press, pp. 89–99.

References

Desmond, Adrian (1989). *The politics of evolution*. Chicago: University of Chicago Press.

Di Gregorio, Mario A (1995). "A wolf in sheep's clothing: Carl Gegenbaur, Ernst Haeckel, the vertebral theory of the skull, and the survival of Richard Owen," *Journal of the History of Biology* 28: 247–280.

Dietrich, Michael R. (1995). "Richard Goldschmidt's 'heresies' and the Evolutionary Synthesis," *Journal of the History of Biology* 28: 431–461.

Dobzhansky, Theodosius (1937). *Genetics and the origin of species*. New York: Columbia University Press.

Dobzhansky, Theodosius (1950). "Human diversity and adaptation," *Cold Spring Harbor Symposia on Quantitative Biology* 15: 385–400.

Dobzhansky, Theodosius (1951). *Genetics and the origin of species*. 3rd and revised New York: Columbia University Press.

Dobzhansky, Theodosius (1955). *Evolution, genetics, and man*. New York: Wiley.

Dobzhansky, Theodosius (1970). *Genetics of the evolutionary process*. New York: Columbia University Press.

Eigen, Edward A. (1997). "Overcoming first impressions: George Cuvier's types," *Journal of the History of Biology* 30: 179–209.

Endler, John A. (1986). *Natural selection in the wild*. Princeton, NJ: Princeton University Press.

Endler, John A., and Tracy McLellan (1988). "The processes of evolution: Towards a newer synthesis," *Annual Review of Ecology and Systematics* 19: 395–421.

Ephrussi, Boris (1951). "Remarks on cell heredity." In *Genetics in the twentieth century*. Edited by L. C. Dunn. New York: Macmillan, pp. 241–262.

Ephrussi, Boris (1953). *Nucleo-cytoplasmic relations in micro-organisms*. Oxford, England: Clarendon Press.

Ereshefsky, Marc (2001). *The poverty of the Linnean hierarchy*. Cambridge, England: Cambridge University Press.

Falk, Raphael (1997). "Muller on development," *Theory in Bioscience* 116: 349–366.

Falk, Raphael (2000). "The gene – a concept in tension." In *The concept of the gene in development and evolution*. Edited by Peter Beurton, Raphael Falk, and Hans-Jörg Rheinberger. Cambridge, England: Cambridge University Press, pp. 317–345.

Falk, Raphael, and Sara Schwartz (1993). "Morgan's hypothesis of the genetic control of development," *Genetics* 134: 671–674.

Farber, Paul L. (1976). "The type-concept in zoology during the first half of the nineteenth century," *Journal of the History of Biology* 9: 93–119.

Gayon, Jean (1989). "Critics and criticisms of the Modern Synthesis: The viewpoint of a philosopher," *Evolutionary Biology* 26: 1–49.

Gayon, Jean (1998). *Darwinism's stuggle for survival*. Cambridge, England: Cambridge University Press.

Gayon, Jean (2000). "From measurement to organization: A philosophical scheme for the history of the concept of heredity." In *The concept of the gene in development and evolution*. Edited by Peter Beurton, Raphael Falk, and Hans-Jörg Rheinberger. Cambridge, England: Cambridge University Press, pp. 69–90.

Gee, Henry (2000). *Shaking the tree: Readings from nature in the history of life*. Chicago: University of Chicago Press.

Gegenbaur, Carl (1870). "Die Stellung und Bedeutung der Morphologie," *Gegenbaurs Morphologisches Jahrbuch* 1: 1–19.

Gerhardt, John, and Marc Kirschner (1997). *Cells, embryos, and evolution*. Malden, MA: Blackwell.

Ghiselin, Michael T. (1969). *The triumph of the Darwinian method*. Berkeley, CA: University of California Press.

Ghiselin, Michael T. (1980). "The failure of morphology to assimilate Darwinism." In *The Evolutionary Synthesis*. Edited by Ernst Mayr and William Provine. Cambridge, MA: Harvard University Press, pp. 180–193.

Ghiselin, Michael T. (1997). *Metaphysics and the origin of species*. Albany, NY: State University of New York Press.

Ghiselin, Michael T. (2002). "An autobiographical anatomy," *History and Philosophy of the Life Sciences* 24: 285–291.

Gilbert, Scott (2001). "Ecological developmental biology: Developmental biology meets the real world," *Developmental Biology* 233: 1–12.

Gilbert, Scott F. (1978). "The embryological origins of the gene theory," *Journal of the History of Biology* 11: 307–351.

Gilbert, Scott F. (1980). "Owen's Vertebral Archetype and evolutionary genetics – a Platonic appreciation," *Perspectives in Biology and Medicine* 23: 475–488.

Gilbert, Scott F. (1988). *Developmental Biology*. 2nd ed. Sunderland, MA: Sinauer Associates.

Gilbert, Scott F. (1991). "Induction and the origins of developmental genetics." In *A conceptual history of modern embryology*. Edited by Scott F. Gilbert. New York: Plenum Press, pp. 181–206.

Gilbert, Scott F. (1996). "Enzymatic adaptation and the entrance of molecular biology into embryology." In *The philosophy and history of molecular biology: New perspectives*. Edited by Sahotra Sarkar. Dordrecht: Kluwer Academic, pp. 101–123.

Gilbert, Scott F. (1998). "Bearing crosses: A historiography of genetics," *American Journal of Medical Genetics* 76: 168–182.

Gilbert, Scott F. (2000). "Genes classical and genes developmental: The different uses of genes in evolutionary syntheses." In *Concept of the gene in development and evolution: Historical and epistemological perspectives*. Edited by Peter Buerton, Hans-Jörg Rheinberger, and Raphael Falk. Cambridge, England: Cambridge University Press, pp. 178–192.

Gilbert, Scott F. (2003a). *Developmental biology*. 7th ed. Sunderland, MA: Sinauer Associates.

Gilbert, Scott F. (2003b). "Evo–devo, devo–evo, and devgen–popgen," *Biology and Philosophy* 18: 347–352.

Gilbert, Scott F., and Jessica A. Bolker (2001). "Homologies of process and modular elements of embryolic construction," *Journal of Experimental Zoology (Molecular and Developmental Evolution)* 291: 1–12.

Gilbert, Scott F., and Marion Faber (1996). "Looking at embryos: The visual and conceptual aesthetics of emerging form." In *The elusive synthesis: Aesthetics and science*. Edited by A. I. Tauber. Dordrecht: Kluwer Academic, pp. 125–151.

Gilbert, Scott F., John M. Opitz, and Rudolf A. Raff (1996). "Resynthesizing evolutionary and developmental biology," *Developmental Biology* 173: 357–372.

References

Gillespie, Charles Coulston (1960). *The edge of objectivity*. Princeton, NJ: Princeton University Press.

Gillespie, Neal C. (1987). "Natural order, natural theology, and social order: John Ray and the 'Newtonian ideology,'" *Journal of the History of Biology* 20: 1–47.

Glacken, Clarence J. (1967). *Traces on the Rhodian shore*. Berkeley, CA: University of California Press.

Goldschmidt, Richard (1940). *The material basis of evolution*. New Haven, CT: Yale University Press.

Gould, Stephen Jay (1977). *Ontogeny and phylogeny*. Cambridge, MA: Harvard University Press.

Gould, Stephen Jay (1980). "G. G. Simpson, paleontology, and the Modern Synthesis." In *The Evolutionary Synthesis*. Edited by Ernst Mayr and William Provine. Cambridge, MA: Harvard University Press, pp. 153–172.

Gould, Stephen Jay (1983). "The hardening of the modern synthesis." In *dimensions of Darwinism*. Edited by Marjorie Grene. Cambridge, England: Cambridge University Press, pp. 71–93.

Gould, Stephen Jay (2002). *The structure of evolutionary theory*. Cambridge, MA: Harvard University Press.

Gould, Stephen Jay, and Richard C. Lewontin (1979). "The spandrels of San Marco and the Panglossian paradigm: A critique of the adaptationist programme," *Proceedings of the Royal Society of London* B205: 581–598.

Gould, Stephen Jay, and Elizabeth S. Vrba (1982). "Exaptation – a missing term in the science of form," *Paleobiology* 8: 4–15.

Grene, John, and Michael Ruse (1994). "Special issue on Ernst Mayr at ninety," *Biology and Philosophy* 9 (3).

Griesemer, James R. (2000). "Reproduction and the reduction of genetics." In *Concept of the gene in development and evolution: Historical and epistemological perspectives*. Edited by Peter Buerton, Hans-Jörg Rheinberger, and Raphael Falk. Cambridge, England: Cambridge University Press, pp. 240–285.

Griesemer, James R., and William C. Wimsatt (1989). "Picturing Weismannism: A case study of conceptual evolution." In *What the philosophy of biology is*. Edited by Michael Ruse. Dordrecht: Kluwer Academic, pp. 75–137.

Griffiths, Paul (1997). *What emotions really are*. Chicago: University of Chicago Press.

Guyer, Paul (1987). *Kant and the claims of knowledge*. Cambridge, England: Cambridge University Press.

Hacking, Ian (1983). *Representing and intervening*. Cambridge, England: Cambridge University Press.

Hall, Brian K. (1994). *Homology: The hierarchical basis of comparative biology*. San Diego, CA: Academic Press.

Hall, Brian K. (1999a). *Evolutionary developmental biology*. 2nd ed. Dordrecht: Kluwer.

Hall, Brian K. (1999b). *Homology*. Chichester, England: Wiley.

Hall, Brian K. (2000). "Evo–devo or devo–evo – does it matter?," *Evolution and Development* 2: 177–178.

Hamburger, Viktor (1980). "Embryology and the Modern Synthesis in evolutionary theory." In *The Evolutionary Synthesis*. Edited by Ernst Mayr and William Provine. Cambridge, MA: Harvard University Press, pp. 97–112.

Harrison, Ross G. (1937). "Embryology and its relations," *Science* 85: 369–374.

Hertwig, Oscar (1894). *The biological problem of to-day: Preformation or epigenesis?* New York: Macmillan.

Hertwig, Paula (1934). "Probleme der heutigen Vererbungslehre," *Die Naturwissenschaften* 25: 425–430.

Hinchliffe, J. Richard (1989). "Reconstructing the archetype: Innovation and conservatism in the evolution and development of the pentadactyl limb." In *Complex organismal functions: Integration and evolution in vertebrates*. Edited by David B. Wake and G. Roth. Chichester, England: Wiley, pp. 171–189.

Hodge, M. J. S. (1985). "Darwin as a lifelong generation theorist." In *The Darwinian heritage*. Edited by David Kohn. Princeton, NJ: Princeton University Press, pp. 207–243.

Holder, Nigel (1983). "Developmental constraints and the evolution of vertebrate digit patterns," *Journal of Theoretical Biology* 104: 451–471.

Holland, P. W. H. (1998). "Major transitions in animal evolution: A developmental genetic perspective," *American Zoologist* 38: 829–842.

Hopkins, William (1860). "Physical theories of the phenomenon of life," *Fraser's Magazine* (June): 739–753.

Horder, T. J. (1989). "Syllabus for an embryological synthesis." In *Complex organismal functions: Integration and evolution in vertebrates*. Edited by David B. Wake and G. Roth. Chichester, England: Wiley, pp. 315–348.

Horn, H. S., J.T. Bonner, W. Pohle, M.J. Katz, M.A.R. Koehl, H. Meinhardt, R.A. Raff, W.-E. Reif, S. C. Stearns, and R. Strathman (1982). "Adaptive aspects of development." In *Evolution and development*. Edited by John Tyler Bonner. New York: Springer-Verlag, pp. 215–235.

Hubby, J. L., and R. C. Lewontin (1966). "A molecular approach to the study of genic heterozygosity in natural populations I. The number of alleles at different loci in *Drosophila pseudoobscura*," *Genetics* 54: 577–594.

Hull, David L. (1965). "The effect of essentialism on taxonomy: 2000 years of stasis," *British Journal for the Philosophy of Science* 15, 16: 314–326, 1–18.

Hull, David L. (1973). *Darwin and his critics*. Chicago: University of Chicago Press.

Hull, David L. (1983). "Darwin and the nature of science." In *Evolution from molecules to men*. Edited by D. S. Bendall. Cambridge, England: Cambridge University Press, pp. 63–80.

Hull, David L. (1999). "Why did Darwin fail? The role of John Stuart Mill." In *Biology and epistemology*. Edited by Richard Creath and Jane Maienschein. Cambridge, England: Cambridge University Press, pp. 48–63.

Hurlbutt, R. H. (1965). *Hume, Newton, and the design argument*. Lincoln, NE: University of Nebraska Press.

Huxley, Julian (1940). *The new systematics*. Oxford, England: Clarendon Press.

Huxley, Julian (1942). *Evolution: The modern synthesis*. New York: Harper and Brother.

Huxley, Thomas Henry (1893a). "Evolution in biology." In *Collected essays vol. 2: Darwiniana*. London: Macmillan, pp. 187–226.

Huxley, Thomas Henry (1893b). "Lectures on evolution." In *Collected essays vol. 4: Science and Hebrew tradition*. London: Macmillan, pp. 46–138.

Huxley, Thomas Henry (1894). "Owen's position in the history of anatomical science." In *The life of Richard Owen, by his grandson*. Vol. 2. Edited by Rev. Richard Owen. London: John Murray 273–332.

Jacob, François (1976). *The logic of life*. New York: Pantheon.

Jacob, François, and Jacques Monod (1961). "Genetic regulatory mechanisms in the synthesis of proteins," *Journal of Molecular Biology* 3: 318–356.

Jacobsen, A. G. (1966). "Inductive processes in embryonic development," *Science* 152 (3718): 25–34.

Jepsen, Glenn L., Ernst Mayr, and George Gaylord Simpson (1949). *Genetics, Paleontology, and evolution*. Princeton, NJ: Princeton University Press.

Johnston, Timothy D., and Gilbert Gottlieb (1990). "Neophenogenesis: A developmental theory of phenotype evolution," *Journal of Theoretical Biology* 147: 471–495.

Keller, Evelyn Fox (2000a). *Century of the gene*. Cambridge, MA: Harvard University Press.

Keller, Evelyn Fox (2000b). "Decoding the genetic program." In *The concept of the gene in development and evolution*. Edited by Peter Beurton, Raphael Falk, and Hans-Jörg Rheinberger. Cambridge, England: Cambridge University Press, pp. 159–177.

Keyes, David N., David L. Lewis, Jane E. Selegue, Bret J. Pearson, Lisa V. Goodrich, Ronald L. Johnson, Julie Gates, Matthew P. Scott, and Sean B. Carrol (1999). "Recruitment of a *hedgehog* regulatory circuit in butterfly eyespot evolution," *Science* 283: 532–534.

Kitcher, Patricia (1990). *Kant's transcendental psychology*. New York: Oxford University Press.

Kohler, Robert E. (1993). "Subcultures in genetics," *Science* 261: 1061–1062.

Kohler, Robert E. (1994). *Lords of the fly*. Chicago: University of Chicago Press.

Lankester, E. Ray (1870). "On the use of the term homology in modern zoology, and the distinction between homogenetic and homoplastic agreements," *Annual Magazine of Natural History* 6: 34–43.

Lankester, E. Ray (1967). "Degeneration: A chapter in Darwinism." In *The interpretation of animal form*. Edited by William Coleman. New York: Johnson Reprint Corporation, pp. 57–132.

Larson, James L. (1994). *Interpreting nature: The science of living form from Linnaeus to Kant*. Baltimore: Johns Hopkins University Press.

Laudan, Larry (1980). "Why was the logic of discovery abandoned?" In *Scientific discovery: Logic and rationality*. Edited by Thomas Nickles. Dordrecht: Reidel, pp. 173–183.

Lenoir, Timothy (1982). *The strategy of life*. Chicago: University of Chicago Press.

Lillie, Frank R. (1927). "The gene and the ontogenetic process," *Science* 66: 361–368.

Love, Alan C. (2003). "Evolutionary morphology, innovation, and the synthesis of evolutionary and developmental biology," *Biology and Philosophy* 18: 309–345.

Love, Alan C., and Rudolf A. Raff (2003). "Knowing your ancestors: Themes in the history of evo–devo," *Evolution and Development* 5: 327–330.

Lovejoy, Arthur O. (1936). *The great chain of being*. Cambridge, MA: Harvard University Press.

Lyell, Charles (1832). *Principles of geology. 1st ed*. London: John Murray.

Lyell, Charles (1835). *Principles of geology. 4th ed*. London: John Murray.

Lyons, Sherrie (1999). *Thomas Henry Huxley: The evolution of a scientist.* Amherst, NY: Prometheus Books.

M'Cosh, James, and George Dickie (1855). *Typical forms and special ends in creation.* New York: Hurst & Company.

Maienschein, Jane (1991a). "The origins of Entwicklungsmechanik." In *A conceptual history of modern embryology.* Edited by Scott F. Gilbert. New York: Plenum Press, pp. 43–61.

Maienschein, Jane (1991b). *Transforming traditions in American biology: 1880–1915.* Baltimore: Johns Hopkins Press.

Maienschein, Jane (1999). "Competing epistemologies and developmental biology." In *Biology and epistemology.* Edited by Richard Creath and Jane Maienschein. Cambridge, England: Cambridge University Press, pp. 122–137.

Maynard Smith, John (1982). *Evolution and the theory of games.* Cambridge, England: Cambridge University Press.

Maynard Smith, John, Richard M. Burian, Stuart A. Kauffman, Pere Alberch, J. Campbell, Brian C. Goodwin, R. Lande, David M. Raup, and Lewis Wolpert (1985). "Developmental constraints and evolution," *Quarterly Review of Biology* 60: 265–287.

Mayr, Ernst (1959a). "Concerning a new biography of Charles Darwin, and its scientific shortcomings: Review of Gertrude Himmelfarb, Darwin and the Darwinian Revolution," *Scientific American* 201 (11): 209–216.

Mayr, Ernst (1959b). "Darwin and the evolutionary theory in biology." In *Evolution and anthropology: A centennial appraisal.* Edited by B. J. Meggers. Washington, DC: Anthropological Society of Washington, pp. 1–10.

Mayr, Ernst (1959c). "Where are we?," *Cold Spring Harbor Symposia on Quantitative Biology* 24: 1–14.

Mayr, Ernst (1961). "Cause and effect in biology," *Science* 134: 1501–1506.

Mayr, Ernst (1966). *Animal species and evolution.* Cambridge, MA: Harvard University Press.

Mayr, Ernst (1970). *Populations, species, and evolution.* Cambridge, MA: Harvard University Press.

Mayr, Ernst (1974). "Behavior programs and evolutionary strategies," *American Scientist* 62: 650–659.

Mayr, Ernst (1976). "Agassiz, Darwin, and evolution." In *Evolution and the diversity of life: Selected essays.* Edited by Ernst Mayr. Cambridge, MA: Harvard University Press, pp. 251–276.

Mayr, Ernst (1980a). "Prologue: Some thoughts on the history of the Evolutionary Synthesis." In *The Evolutionary Synthesis.* Edited by Ernst Mayr and William Provine. Cambridge, MA: Harvard University Press, pp. 1–48.

Mayr, Ernst (1982). *The growth of biological thought.* Cambridge, MA: Harvard University Press.

Mayr, Ernst (1984). "The triumph of the evolutionary synthesis," *Times Literary Supplement November 2, 1984*: 1261–1262.

Mayr, Ernst (1988). *Toward a new philosophy of biology.* Cambridge, MA: Harvard University Press.

Mayr, Ernst (1991). "An overview of current evolutionary biology." In *New perspectives on evolution.* Edited by Leonard Warren and Hilary Koprowski. New York: Wiley Liss, pp. 1–14.

References

Mayr, Ernst (1992). "Controversies in retrospect," *Oxford Surveys in Evolutionary Biology* 8: 1–34.

Mayr, Ernst (1994a). "Recapitulation reinterpreted: The somatic program," *Quarterly Review of Biology* 69: 223–232.

Mayr, Ernst (1994b). "Reply to John Beatty," *Biology and Philosophy* 9: 357–358.

Mayr, Ernst and William Provine (1980). *The Evolutionary Synthesis.* Cambridge, MA: Harvard University Press.

McOuat, Gordon R. (1996). "Species, rules and meaning: The politics of language and the ends of definitions in 19th century natural history," *Studies in History and Philosophy of Science Part A* 27 (4): 473–519.

Mivart, St. George (1871). *On the genesis of species.* New York: Appleton.

Moore, James (1991). "Deconstructing Darwinism: The politics of evolution in the 1860s," *Journal of the History of Biology* 24: 353–408.

Morange, Michel (1998). *A history of molecular biology.* Cambridge, MA: Harvard University Press.

Morgan, Thomas Hunt (1910). "Chromosomes and heredity," *American Naturalist* 44: 449–496.

Morgan, Thomas Hunt (1919). *The physical basis of heredity.* Philadelphia: Lippincott.

Morgan, Thomas Hunt (1926a). *The theory of the gene.* New Haven, CT: Yale University Press.

Morgan, Thomas Hunt (1926b), "Genetics and the physiology of development," *American Naturalist* 60: 489–515.

Morgan, Thomas Hunt (1932). "The rise of genetics," *Science* 76: 261–267, 285–288.

Morgan, Thomas Hunt, A. H. Sturtevant, H. J. Muller, and C. B. Bridges (1915). *The mechanism of Mendelian heredity.* New York: Henry Holt.

Multauf, Robert P. (1966). *The origins of chemistry.* London: Oldbourne.

Müller, Fritz (1869). *Facts and arguments for Darwin.* London: John Murray.

Müller-Wille, Staffan (1995). "Linnaeus concept of a "symmetry of all parts," *Jahrbuch für Geschichte und Theorie der Biologie* 2: 41–47.

Nyhart, Lynn (1995). *Biology takes form.* Chicago: University of Chicago Press.

Oken, Lorenz (1807). *Über die Bedeutrung der Schadelknochen.* Jena, Germany: Göpferdt.

Olson, Everett C. (1960). "Morphology, paleontology, and evolution." In *Evolution after Darwin, vol. 1: The evolution of life.* Edited by Sol Tax. Chicago: Chicago University Press, pp. 523–545.

Oppenheimer, Jane M. (1966). "The growth and development of developmental biology." In *Major problems in developmental biology.* Edited by Michael Locke. New York: Academic Press, pp. 1–27.

Ospovat, Dov (1981). *The development of Darwin's theory.* Cambridge, England: Cambridge University Press.

Owen, Richard (1843). *Lectures on the comparative anatomy and physiology of the invertebrate animals.* London: Longman Brown Green and Longmans.

Owen, Richard (1848). *The archetype and homologies of the vertebrate skeleton.* London: J. van Voorst.

Owen, Richard (1849). *On the nature of limbs.* London: J. van Voorst.

Paley, William (1809). *Natural theology.* 12th ed. London: J. Faulder.

Pearson, Karl (1892). *Grammar of science.* London: Walter Scott.

Popper, Karl R. (1950). *The open society and its enemies*. Princeton, NJ: Princeton University Press.

Poulton, Edward B. (1908). "What is a species?" In *Essays on evolution*. Edited by Edward B. Poulton. Oxford, England: Clarendon Press, pp. 46–94.

Provine, William (1971). *The origins of theoretical population genetics*. Chicago: University of Chicago Press.

Provine, William (1980). "Epilogue." In *The Evolutionary Synthesis*. Edited by Ernst Mayr and William Provine. Cambridge, MA: Harvard University Press, pp. 399–411.

Provine, William (1986). *Sewall Wright and evolutionary biology*. Chicago: University of Chicago Press.

Provine, William (1988). "Progress in evolution and meaning in life." In *Evolutionary progress*. Edited by Matthew H. Nitecki. Chicago: University of Chicago Press, pp. 49–74.

Quine, Willard van Orman (1969). "Natural kinds." In *Ontological relativity and other essays*. Edited by Willard van Orman Quine. New York: Columbia University Press, pp. 114–138.

Raff, Rudolf, and Thomas C. Kaufmann (1983). *Embryos, genes, and evolution*. Bloomington, IN: Indiana University Press.

Raff, Rudolf A. (1996). *The shape of life*. Chicago: University of Chicago Press.

Raven, Charles E. (1953). *Science and religion*. Cambridge, England: Cambridge University Press.

Reeve, Hudson Kern, and Paul W. Sherman (1993). "Adaptation and the goals of evolutionary research," *Quarterly Review of Biology* 68: 1–32.

Rehbock, Philip F. (1985). *The philosophical naturalists*. Madison, WI: University of Wisconsin Press.

Reif, Wolf-Ernst, Thomas Junker, and Uwe Hossfeld (2000). "The synthetic theory of evolution: General problems and the German contribution to the synthesis," *Theory in Biosciences* 119: 41–91.

Richards, Eveleen (1987). "A question of property rights: Richard Owen's evolutionism reassessed," *British Journal for the History of Science* 20: 129–171.

Richards, Robert J. (2002). *The Romantic conception of life*. Chicago: University of Chicago Press.

Rienesl, J., and Gunter P. Wagner (1992). "Constancy and change of basipodial variation patterns: A comparative study of crested and marbled newts – *Triturus cristatus, Triturus marmoratus* – and their natural hybrids," *Journal of Evolutionary Biology* 5: 307–324.

Robert, Jason Scott (2002). "How developmental is evolutionary developmental biology?," *Biology and Philosophy* 17: 591–611.

Roe, Shirley (1981). *Matter, life, and generation*. Cambridge, England: Cambridge University Press.

Roget, Peter Mark (1834). *Animal and vegetable physiology considered with respect to natural theology. 2 vols*. London: William Pickering.

Roth, V. Louise (1984). "On homology," *Biological Journal of the Linnean Society* 22: 13–29.

Roth, V. Louise (1988). "The biological basis of homology." In *Ontogeny and systematics*. Edited by C. J. Humphries. New York: Columbia University Press, pp. 1–26.

References

Roux, Wilhelm (1986). "The problems, methods, and scope of developmental mechanics." In *Defining biology: Lectures from the 1890s*. Edited by Jane Maienschein. Cambridge, MA: Harvard University Press, pp. 108–146.

Rupke, Nicolaas A. (1993). "Richard Owen's vertebrate archetype," *ISIS* 84: 231–251.

Rupke, Nikolaas A. (1994). *Richard Owen: Victorian naturalist*. New Haven, CT: Yale University Press.

Ruse, Michael (1979). *The Darwinian revolution*. Chicago: University of Chicago Press.

Ruse, Michael (1996). *Monad to man: The concept of progress in evolutionary biology*. Cambridge, MA: Harvard University Press.

Russell, Edwin S. (1916). *Form and function*. London: John Murray.

Saha, Margaret (1991). "Spemann seen through a lens." In *A conceptual history of modern embryology*. Edited by Scott F. Gilbert. New York: Plenum Press, pp. 91–108.

Sander, Klaus (1983). "The evolution of patterning mechanisms: Gleanings from insect embryogenesis and spermatogenesis." In *Development and evolution*. Edited by Brian C. Goodwin, Nigel Holder, and C. C. Wylie. Cambridge, England: Cambridge University Press, pp. 353–379.

Sander, Klaus (1985). "The role of genes in ontogenesis: Evolving concepts from 1883 to 1983 as perceived by an insect embryologist." In *A history of embryology*. Edited by T. J. Horder, J. A. Witkowski, and C. C. Wylie. Cambridge, England: Cambridge University Press, pp. 363–395.

Sandler, Iris, and Laurence Sandler (1985). "A conceptual ambiguity that contributed to the neglect of Mendel's paper," *History and Philosophy of the Life Sciences* 7: 3–70.

Sapp, Jan (1987). *Beyond the gene: Cytoplasmic inheritance and the struggle for authority in genetics*. New York: Oxford University Press.

Sapp, Jan (1990). "The nine lives of Gregor Mendel." In *Experimental inquiries*. Edited by H. E. Le Grand. Dordrecht: Kluwer Academic, pp. 137–166.

Sarkar, Sahotra (1999). "From the *Reactionsnorm* to the adaptive norm: The norm of reaction, 1909–1960," *Biology and Philosophy* 14: 235–252.

Scharf, Sara (2003). "Contrasts along a continuum: Purposes and constraints in natural and artificial classifications in systematic biology, 1730s–1830s," unpublished manuscript, University of Toronto.

Schwartz, Sara (2000). "The differential concept of the gene." In *Concept of the gene in development and evolution: Historical and epistemological perspectives*. Edited by Peter Buerton, Hans-Jörg Rheinberger, and Raphael Falk. Cambridge, England: Cambridge University Press, pp. 26–39.

Schwenk, Kurt (1995). "A utilitarian approach to evolutionary constraint," *Zoology* 98: 251–262.

Shubin, Neil, Cliff Tabin, and Sean Carrol (1997). "Fossils, genes and the evolution of animal limbs," *Nature* 388: 639–648.

Shubin, Neil, Cliff Tabin, and Sean Carroll (2000). "Fossils, genes and the evolution of animal limbs." In *Shaking the tree: Readings from nature in the history of life*. Edited by Henry Gee. Chicago: University of Chicago Press, pp. 100–127.

Shubin, Neil, David B. Wake, and Andrew J. Crawford (1995). "Morphological variation in the limbs of *Taricha granulosa* (Caudata: Salamandridae): Evolutionary and phylogenetic mechanisms," *Evolution* 49: 874–884.

Shubin, Neil H., and Pere Alberch (1986). "A morphological approach to the origin and basic organization of the tetrapod limb," *Evolutionary Biology* 20: 319–387.

Sloan, Phillip R. (1992). "On the edge of evolution." In *The Hunterian lectures in comparative anatomy, May and June 1837.* Edited by Richard Owen. Chicago: University of Chicago Press, pp. 3–72.

Sloan, Phillip R. (2002). "Preforming the categories: Eighteenth-century generation theory and the biological roots of Kant's a priori," *Journal of the History of Philosophy* 40: 229–253.

Sloan, Phillip R. (2003). "Whewell's philosophy of discovery and the Archetype of the vertebrate skeleton," *Annals of Science* 60: 39–61.

Smith, W. C., and R. M. Harland (1992). "Expression cloning of *noggin*, a new dorsalizing factor localized to the Spemann organizer in *Xenopus* embryos," *Cell* 70: 829–840.

Smocovitis, Vassiliki Betty (1992). "Unifying biology: The Evolutionary Synthesis and modern biology," *Journal of the History of Biology* 25: 1–65.

Smocovitis, Vassiliki Betty (1996). *Unifying biology: The Evolutionary Synthesis and modern biology.* Princeton, NJ: Princeton University Press.

Smocovitis, Vassiliki Betty (1999). "The 1959 Darwin centennial celebration in America," *Osiris* 14: 274–323.

Sober, Elliott (2000). *Philosophy of biology.* 2nd. ed. Boulder, CO: Westview Press.

Spemann, H. (1915). "Zur Geschichte und Kritik des Begriffs der Homologie." In *Die Kultur der Gegenwart.* Edited by E. Hinneberg. Leipzig, Germany: Teubner, pp. 63–96.

Spurway, H. (1949). "Remarks on Vavilov's law of homologous variation," *Supplemento a La Ricerca Scientifica* 19: 18–24.

Stevens, Peter F. (1994). *The development of biological systematics.* New York: Columbia University Press.

Stevens, Peter F. (1997). "How to interpret botanical classifications – suggestions from history," *Bioscience* 47: 243–250.

Stevens, Peter F., and S. P. Cullens (1990). "Linnaeus, the cortex-medulla theory, and the key to his understanding of plant form and natural relationships," *Journal of the Arnold Arboretum* 71: 179–220.

Strickland, Hugh E. (1840). "Observations on the affinities and analogies of organized beings," *Magazine of Natural History* 4: 210–226.

Tax, Sol (1960). *Evolution after Darwin vol. 1: The evolution of life.* Chicago: Chicago University Press.

Thieffry, Denis (1996). "*Escherichia coli* as a model system with which to study cell differentiation," *History and Philosophy of the Life Sciences* 18: 163–193.

van der Steen, Wim J. (1996). "Screening-off and natural selection," *Philosophy of Science* 63: 115–121.

van der Weele, Cor (1999). *Images of development: Environmental causes in ontogeny.* Albany, NY: State University of New York Press.

van Fraassen, Bas (1980). *The scientific image.* Oxford: Clarendon Press.

van Speybroeck, Linda (2002). "Philosophers and biologists exploring epigenetics," *Biology and Philosophy* 17: 743–746.

Vavilov, N. I. (1922). "The law of homologous series in variation," *Journal of Genetics* 12: 47–87.

References

Waddington, Conrad H. (1939). *An introduction to modern genetics*. London: Allen & Unwin.

Waddington, Conrad H. (1940). *Organisers and genes*. London: Cambridge University Press.

Waddington, Conrad H. (1953). "Epigenetics and evolution." In *Symposia of the Society for Experimental Biology VII: Evolution*. Cambridge, England: Cambridge University Press, pp. 186–199.

Wagner, Günter P. (1989a). "The biological homology concept," *Annual Review of Ecology and Systematics* 20: 51–69.

Wagner, Günter P. (1989b). "The origin of morphological characters and the biological basis of homology," *Evolution* 43: 1157–1171.

Wagner, Günter P. (2000). "What is the promise of developmental evolution? Part I: Why is developmental biology necessary to explain evolutionary innovations," *Journal of Experimental Zoology* 288: 95–98.

Wagner, Günter P. (2001a). *The character concept in evolutionary biology*. San Diego, CA: Academic Press.

Wagner, Günter P. (2001b). "Characters, units, and natural kinds: An introduction." In *The character concept in evolutionary biology*. Edited by Günter P. Wagner. San Diego, CA: Academic Press, pp. 1–10.

Wagner, Günter P. (forthcoming). *The biological homology concept and its applications*. Princeton, NJ: Princeton University Press.

Waisbren, Steven James (1988). "The importance of morphology in the Evolutionary Synthesis as demonstrated by the contributions of the Oxford Group: Goodrich, Huxley, and de Beer," *Journal of the History of Biology* 21: 291–330.

Wake, David B. (1991). "Homoplasy: The result of natural selection, or evidence of design limitations?," *American Naturalist* 138: 543–567.

Wake, David B. (1996). "Evolutionary developmental biology – prospects for an evolutionary synthesis at the developmental level." In *New perspectives on the history of life: Essays on systematic biology as historical narrative*. Edited by Michael T. Ghiselin and G. Pinna. San Francisco: California Academy of Sciences, pp. 97–107.

Wallace, Bruce (1986). "Can embryologists contribute to an understanding of evolutionary mechanisms?" In *Integrating scientific disciplines*. Edited by William Bechtel. Dordrecht: Martinus Nijhoff, pp. 149–163.

Walsh, Denis M. (2003). "Fit and diversity: Explaining adaptive evolution," *Philosophy of Science* 70: 280–301.

Waterhouse, George R. (1843). "Observations on the classification of the Mammalia," *Annals and Magazine of Natural History* 12: 399–412.

Whewell, William (1836). *Astronomy and general physics, considered with reference to natural theology*. Philadelphia: Carey, Lea & Blanchard.

Whewell, William (1837). *History of the inductive sciences: From the earliest to the present times*. 2 vols., 1st ed. New York: Appleton.

Whewell, William (1863). *History of the inductive sciences: From the earliest to the present times*. 2 vols., 3rd ed., with additions. New York: Appleton.

Williams, George C. (1966). *Adaptation and natural selection*. Princeton, NJ: Princeton University Press.

Williams, George C. (1992). *Natural selection: Domains, levels, and challenges*. Oxford, England: Oxford University Press.

References

Winsor, Mary P. (1976). *Starfish, jellyfish, and the order of life*. New Haven, CT: Yale University Press.

Winsor, Mary P. (1979). "Louis Agassiz and the species question," *Studies in History of Biology* 4: 89–117.

Winsor, Mary P. (2003). "Non-essentialist methods in pre-Darwinian taxonomy," *Biology and Philosophy* 18: 387–400.

Winsor, Mary P. (forthcoming). "The creation of the Essentialism Story (an exercise in metahistory)," *Journal of the History of Biology*.

Winther, Rasmus G. (2000). "Darwin on variation and heredity," *Journal of the History of Biology* 33: 425–455.

Winther, Rasmus G. (2001). "August Weismann on germ-plasm variation," *Journal of the History of Biology* 34: 517–555.

Wolpert, Lewis (1991). *The triumph of the embryo*. Oxford, England: Oxford University Press.

Wray, Gregory A. (2001). "Development: Resolving the *Hox* paradox," *Science* 292 (5525): 2256–2257.

Wright, Sewall (1941). "The physiology of the gene," *Physiological Reviews* 21: 487–527.

Wright, Sewall (1945). "Genes as physiological agents," *American Naturalist* 79: 289–303.

Zirkle, Conway (1951). "The knowledge of heredity before 1900." In *Genetics in the 20th century*. Edited by Leslie C. Dunn. New York: Macmillan, pp. 35–57.

Zirkle, Conway (1959). "Species before Darwin," *Proceedings of the American Philosophical Society* 103: 636–644.

Index